Pesticide residues in food — 1987

Report of the Joint Meeting of the
FAO Panel of Experts on Pesticide Residues
in Food and the Environment
and a WHO Expert Group on Pesticide Residues
Geneva, 21-30 September 1987

FAO
PLANT
PRODUCTION
AND PROTECTION
PAPER

84

FOOD
AND
AGRICULTURE
ORGANIZATION
OF THE
UNITED NATIONS
Rome, 1987

Monographs containing summaries of residue data and toxicological data considered at the 1986 JMPR, together with recommendations, are available upon request from FAO under the title:

Pesticide residues in food - 1987
Evaluations 1987
Part I: Residues
Part II: Toxicology
FAO Plant Production and Protection Paper

The designations employed and the presentation of material in this publication do not imply the expression of any opinion whatsoever on the part of the Food and Agriculture Organization of the United Nations concerning the legal status of any country, territory, city or area or of its authorities, or concerning the delimitation of its frontiers or boundaries.

This report contains the collective views of two international groups of experts and does not necessarily represent the decisions or the stated policy of the Food and Agriculture Organization of the United Nations or of the World Health Organization.

INTERNATIONAL PROGRAMME ON CHEMICAL SAFETY

The preparatory work for the toxicological evaluations of pesticide residues carried out by the WHO Expert Group on Pesticide Residues for consideration by the FAO/WHO Joint Meeting on Pesticide Residues in Food and the Environment is actively supported by the International Programme on Chemical Safety (IPCS).
The International Programme on Chemical Safety (IPCS) is a joint venture of the United Nations Environment Programme, the International Labour Organisation, and the World Health Organization. One of the main objectives of the IPCS is to carry out and disseminate evaluations of the effects of chemicals on human health and the quality of the environment.

M-84
ISBN 92-5-102620-3

TABLE OF CONTENTS

---

* First toxicological evaluation

** First full evaluation

- v -

1987 JOINT MEETING OF THE FAO PANEL OF EXPERTS ON PESTICIDE RESIDUES
IN FOOD AND THE ENVIRONMENT AND THE WHO EXPERT GROUP ON
PESTICIDE RESIDUES

Geneva, 21-30 September 1987

<u>FAO Panel of Experts on Pesticide Residues in Food and the Environment</u>

Dr D.C. Abbott, formerly Deputy Director (Government Analyst), Laboratory of
the Government Chemist, London, UK (Vice-Chairman)

Professor Dr A.F.H. Besemer, formerly Chair of Phytopharmacy, Agricultural
University, Wageningen, The Netherlands

Dr P.F. Dos Santos, Head, National Research Centre for Plant Protection
(EMBRAPA) Jaguariuna-SP, Brazil

Dr Roy Greenhalgh, Plant Research Centre, Agriculture Canada, Ottawa, Ontario,
K1A OC6 Canada

Mr D.J. Hamilton, Assistant Director Agricultural Chemistry Branch, Department
of Primary Industries, Brisbane, Australia (Rapporteur)

Dr E.D. Magallona, Head, Pesticide Toxicology and Chemistry Laboratory,
National Crop Protection Center, College of Agriculture, University
of the Philippines at Los Baños, College, Laguna, Philippines

<u>WHO Expert Group on Pesticide Residues</u>

Professor Mrs Nabila M.S. Bakry, Head Professor of Pesticide Chemistry and
Toxicology, Department of Plant Protection, Faculty of Agriculture,
University of Alexandria, El-Shatby, Alexandria, Egypt

Dr V. Benes, Chief, Department of Toxicology, Institute of Hygiene and
Epidemiology, Prague, Czechoslovakia

Professor J.F. Borzelleca, Professor of Pharmacology and Toxicology, Medical
College of Virginia, Richmond, VA, USA

Mr D.J. Clegg, Head, Pesticide Section, Toxicological Evaluation Division,
Food Directorate, Health Protection Branch, Ottawa, Ontario, K1A OL2
Canada

Professor M. Lotti, Istituto di Medicina del Lavoro, University of Padua
Medical School, Padua, Italy (Chairman)

Professor F.G. Reyes, Department of Food Science, School of Food
Engineering/UNICAMP, Campinas, Brazil (Rapporteur)

<u>Secretariat</u>

Mr J.A.R. Bates, formerly Head, Pesticide Registration and Surveillance
     Department, Ministry of Agriculture, Fisheries and Food, <u>Harpenden</u>, UK
     (FAO Consultant)

Professor C.L. Berry, Professor of Pathology, Chairman, Department of Morbid
     Anatomy, The London Hospital Medical College, <u>London</u>, England (WHO
     Temporary Adviser)

Dr A.L. Black, Medical Adviser in Toxicology, Commonwealth Department of Health
     Woden, <u>A.C.T.</u>, Australia (WHO Temporary Adviser)

Dr J.R.P. Cabral, Scientist, Unit of Mechanisms of Carcinogenesis,
     International Agency for Research on Cancer, <u>Lyon</u>, France (WHO Temporary
     Adviser)

Dr J.L. Herrman, Food and Drug Administration, <u>Washington, DC</u>, USA (WHO
     Consultant)

*Dr G. Hudson, Head of Division, Directorate-General for Agriculture,
     Commission of the European Communities, <u>Brussels</u>, Belgium (WHO Temporary
     Adviser)

Dr F.-W. Kopisch-Obuch, Pesticide Residue Specialist, Plant Protection Service,
     FAO, <u>Rome</u>, Italy (Joint Secretary)

Dr L.G. Ladomery, Food Standards Officer, Joint FAO/WHO Food Standards
     Programme, FAO, <u>Rome</u>, Italy

Mr A.F. Machin, <u>London</u>, England (FAO Consultant)

Dr M. Manno, Istituto di Medicina del Lavoro, University of Padua Medical
     School, <u>Padua</u>, Italy (WHO Temporary Adviser)

Mr A.J. Pieters, Ministry of Welfare, Health and Cultural Affairs, <u>Rijswijk</u>
     The Netherlands (FAO Consultant)

Dr D.S. Saunders, Hazard Evaluation Division, US Environmental Protection
     Agency, <u>Washington, D.C.</u>, USA (WHO Temporary Adviser)

Dr A. Takanaka, Head, Division of Pharmacology, Biological Safety Research
     Center, National Institute of Hygienic Sciences, <u>Tokyo</u>, Japan (WHO
     Temporary Adviser)

Dr Mrs Engelina M. den Tonkelaar, Toxicology Advisory Center, National
     Institute of Public Health and Environmental Hygiene, <u>Bilthoven</u>,
     The Netherlands (WHO Temporary Adviser)

Dr G. Vettorazzi, Senior Toxicologist, International Programme on Chemical
     Safety, Division of Environmental Health, WHO, <u>Geneva</u>, Switzerland
     (Joint Secretary)

Mr M. Walsh, Directorate General for Agriculture, Commission of the European
     Communities, <u>Brussels</u>, Belgium, (FAO Consultant)

---

* invited but unable to attend

ABBREVIATIONS WHICH MAY BE USED IN THIS REPORT

(n.b.: chemical elements and pesticides are not included in this list)

AChE	acetylcholinesterase
ADI	acceptable daily intake
ai	active ingredient
approx.	approximate
at. wt.	atomic weight
b.p.	boiling point
c	centi - (x 10^{-2})
°C	degree Celsius (centigrade)
CCPR	Codex Committee on Pesticide Residues
cm	centimetre
CNS	central nervous system
cu	cubic
$\underline{DL}$	racemic (optical configuration, a mixture of dextro- and laevo-; preceding a chemical name)
EC	emulsion concentrate
ERL	extraneous residue limit
F_1	filial generation, first
F_2	filial generation, second
f.p.	freezing point
FAO	Food and Agriculture Organization of the United Nations
g	gram
µg	microgram
GAP	good agricultural practice
GC-MS	gas chromatography - mass spectrometry
G.I.	gastro-intestinal
GL	guideline level
GLC	gas-liquid chromatography
GPC	gel-permeation chromatography
h	hour(s)
ha	hectare
Hb	haemoglobin
HPLC	high-performance liquid chromatography
IBT	Industrial Bio-Test Laboratories
i.m.	intramuscular
i.p.	intraperitoneal
IR	infrared
i.v.	intravenous
JMPR	Joint FAO/WHO Meeting on Pesticide Residues (Joint Meeting of the FAO Panel of Experts on Pesticide Residues in Food and the Environment and a WHO Expert Group on Pesticide Residues)

k	kilo- (x 10^3)
kg	kilogram
l	litre
LC_{50}	lethal concentration, 50%
LD_{50}	lethal dose, median
m	metre
mg	milligram
μm	micrometre (micron)
min	minute(s)
ml	millilitre
MLD	minimum lethal dose
mm	millimetre
M	molar
mo	month(s)
m.p.	melting point
MRL	Maximum Residue Limit (This term replaces "tolerance")
MTD	maximum tolerated dose
n	normal (defining isomeric configuration)
NCI	National Cancer Institute (United States)
NMR	nuclear magnetic resonance
no.	number
NOEL	no-observed-effect level
NOAEL	no-observed-adverse-effect level
NTE	neuropathy target esterase
o	ortho (indicating position in a chemical name)
OP	organophosphorus pesticide
p	para (indicating position in a chemical name)
PHI	pre-harvest interval
ppm	parts per million (Used only reference to the concentration of a pesticide in an experimental diet. In all other contexts the terms mg/kg or mg/l are used).
s.c.	subcutaneous
SD	standard deviation
SE	standard error
sp./spp.	species (only after a generic name)
sp gr	specific gravity
sq	square
t	tonne (metric ton)
TADI	Tempory Acceptable Daily Intake
tert	tertiary (in a chemical name)
TLC	thin-layer chromatography
TMRL	Temporary Maximum Residue Limit
UV	ultraviolet
v/v	volume ratio (volume per volume)

WHO	World Health Organization
wk	week
WP	wettable powder
wt	weight
wt/vol	weight per volume
w/w	weight per weight
yr	year
<	less than
≤	less than or equal to
>	greater than
≥	greater than or equal to

PESTICIDE RESIDUES IN FOOD

Report of the 1987 Joint FAO/WHO Meeting of Experts

1. Introduction

A joint meeting of the FAO Panel of Experts on Pesticide Residues in Food and the Environment and a WHO Expert Group on Pesticide Residues (JMPR) was held in Geneva, Switzerland, from 21-30 September 1987. The meeting was opened by Dr J.-P. Jardel, Assistant Director-General, WHO, on behalf of the Directors-General of FAO and WHO. The FAO Panel and the WHO Expert Group had met in preparatory sessions on 17-18 September.

The meeting was held in pursuance of recommendations made by previous meetings and accepted by the governing bodies of FAO and WHO that studies should be undertaken jointly by experts to evaluate possible hazards to man arising from the occurrence of residues of pesticides in foods. The reports of previous joint meetings (see references, Section 7) contain information on acceptable daily intakes for man (ADIs), maximum residue limits (MRLs) and general principles of evaluation for the various pesticides considered. The supporting documents contain detailed monographs on these pesticides and include comments on analytical methods. The present meeting was convened to consider a further number of pesticides together with items of a general or a specific nature. These include items for clarification of recommendations made at previous meetings or for reconsideration of previous evaluations in the light of findings of subsequent research or other developments.

During the meeting the FAO Panel of Experts was responsible for reviewing pesticide use patterns (good agricultural practices), data on the chemistry and composition of pesticides and methods of analysis of pesticide residues and for estimating the maximum residue levels that might occur as a result of the use of the pesticide according to good agricultural practices. The WHO Expert Group was responsible for reviewing toxicological and related data and for estimating, where possible, ADIs for man for the pesticides. The recommendations of the joint meeting, including further research and information, are proposed for use by Member Governments of the respective agencies and other interested parties.

2. GENERAL CONSIDERATIONS

2.1 MODIFICATIONS TO THE AGENDA

The meeting did not consider fluvalinate owing to a lack of data.

Several questions raised at the 19th Session of the CCPR were considered. These are listed in Section 2.11.

2.2 DATA UTILIZED IN ESTABLISHING ADIs

The Joint Meeting assumes that the data provided to it for determining an ADI are accurate and valid. The responsibility for assuring accuracy and validity (or authentication of reports) rests with those submitting the data. Where possible (when detailed data are available) the JMPR also assesses validity and accuracy. However, published reports (for example NCI reports) and papers from refereed journals must be assumed to be accurate and valid since the detailed data are not usually available for evaluation.

2.3 PURITY OF TEST MATERIALS USED IN TOXICITY STUDIES

On previous occasions (1968, 1969, 1970, 1974, 1977, 1978, 1980, 1981, 1984) the JMPR has drawn attention to the purity of test compounds used in toxicity tests.

The WHO Group of Experts again discussed the problem of the variation in the quantitative and qualitative purity of the test material used in toxicity assays. The Group appreciated that, although the data available to them were usually generated using material from one source, the purity of different production batches could vary.

The Group wishes to reiterate that the ADI is estimated from data derived from tests using compounds of specific purity. While such ADIs can be relevant to a product of different origin or purity, there are examples where the amounts or types of impurities in the technical material can markedly influence the toxicity of and, consequently, the ADI for the compound. Where data are available, such cases will be identified in the monographs. However, adequate data of this type are not often available. Caution is therefore required in applying the ADI to a pesticide whose purity differs significantly from that used in toxicity tests reviewed by the WHO Group.

The Group considered that the purity of the test material should always be specified both qualitatively and quantitatively.

2.4 PRINCIPLES FOR THE SAFETY ASSESSMENT OF PESTICIDE RESIDUES IN FOOD

The meeting noted the report of a planning meeting convened on behalf of the Central Unit of the International Programme on Chemical Safety (IPCS), which was held in response to a recommendation by the 1985 JMPR for an international meeting to consider principles utilized by the WHO Group. The report provided a comprehensive outline of a proposed monograph on an update of the principles for the safety assessment of pesticide residues in food. The meeting was informed that a draft document based on this outline is to be prepared, and should be available by the spring of 1988. The WHO Group expressed the hope that this draft would be available for discussion at the next meeting of the JMPR.

2.5 DOSES IN TOXICITY STUDIES AND EXTRAPOLATION TO MAN

The meeting was concerned at the difficulties of interpretation of the results of long-term studies in which high doses have been used. In reproduction and teratology studies, the use of maternally toxic doses has also caused concern. The meeting discussed the maximum tolerated dose (MTD) which has been defined as "a dose that does not shorten life expectancy nor produce signs of toxicity other than those due to cancer", and "operationally as the maximum dose level at which a substance induces a decrement in weight gain of no greater than 10% in a subchronic toxicity test" (WHO, 1987). To identify agents with potentially low orders of toxicity, exposure conditions are often maximized. These may include the use of very high doses and gavage administration. A number of assumptions are implicit in the use of the MTD: (i) the absorption, distribution, biotransformation and excretion of a chemical are not dose-dependent (that is, their kinetics are the same at low and high doses), (ii) both the rate and extent of reparative processes (for example, DNA repair) are independent of dose and of the extent of damage, (iii) the response to a chemical is not age-dependent, (iv) the dose-dependent response is linear, (v) doses tested in animals need not bear any relationship to human exposure levels.

Data on drugs and non-drug chemicals indicate that the processes involved in absorption, distribution, biotransformation and excretion are dependent upon many factors, including physico-chemical properties, extent of protein-binding, bioavailability and dose. Some of these processes are saturable. Products of biotransformation may be formed at different rates and in different quantities or by different pathways at high doses (for example, 2-phenylphenol). Comparative biotransformation data should be obtained early in an evaluation programme to identify the species that closely resemble man (cf WHO, 1987). Since human data may not be available, in vitro data from multiple species, including human tissue or cultured human cells, would identify the appropriate species that handles the chemical via the same pathways as man. It is valid to extrapolate animal data to man only if the biotransformation pathways of the chemical are identical or very similar between species and if the doses do not exceed the capacity of the pathways being compared. If this capacity is exceeded, different metabolites may be produced.

Kinetic data are useful in the design of studies and in the interpretation and extrapolation of the data. For example, if the test material is not absorbed, the need for one or more long-term studies would be obviated.

Extrapolation of animal data to man may be compromised by differences in the movement of the chemical after absorption. For example, the administration of high doses of certain chemicals may result in increased enterohepatic circulation of the chemical and/or its metabolites. This is an important system in the rat but less so in man.

In reproductive studies, maternally toxic doses producing severe effects (indicated by loss of weight, or failure of maternal weight gain) almost invariably lead to foetal growth retardation which predisposes to structural abnormalities (as happens in diabetes mellitus in man). This growth retardation may act as a confounding variable; in the presence of excessive maternal toxicity, definitive statements about teratology cannot be made. The proper selection of doses in reproduction studies is addressed in a recent IPCS publication (WHO, 1987).

The use of the MTD assumes a linear dose-response curve (the effect produced is qualitatively similar at every dose), an approach which is presumed to be conservative. Responses to a chemical may vary with increasing doses, as exemplified by the organophosphorus pesticides, where the primary response at relatively low doses is cholinesterase inhibition whereas at high doses delayed neurotoxicity may be induced. For carcinogenesis, this assumption has been shown to be false in the "ED01" or mega-mouse study (involving 24,000 mice carried out at the National Center for Toxicological Research, USA) using the compound 2-acetylaminofluorene. More animals were exposed at low than at high doses so that low level effects could be detected with confidence. For neither bladder nor liver tumours was the dose response linear. The same problem of non-linearity has been demonstrated for formaldehyde, saccharin and a number of other carcinogens.

Properly designed biotransformation studies over a range of doses including human exposure levels where available would provide a rational basis for doses to be used in chronic animal studies.

Human data should be an integral part of the material evaluated in any regulatory assessment and may be valuable in the interpretation of the results of long-term animal tests. Data on human exposure should always be collected as part of the continuing evaluation of the safety of compounds in commercial use. Sources of these data include: workers involved in the production of the chemical, workers involved in the application (use) of the chemical, subjects exposed to the test chemical under carefully defined and controlled experimental conditions and consumer exposure data. The results of these studies could then be used in the design of further toxicological work.

The interpretation of toxicity data obtained in animals and its extrapolation to humans is a complex process. Only accurate and validated data from properly designed (appropriate species, doses and exposure conditions) and executed studies should be used. The proper design of definitive long-term studies should be based on comparative data on the absorption, distribution, biotransformation, excretion and appropriate kinetic considerations of the test substance. Data obtained from studies involving the use of very high doses (for example, the MTD) or unusual exposure conditions (for example, gavage) should be carefully interpreted in the light of the available kinetic findings. Evaluation of the safety of chemicals in commerce is a dynamic process which recognizes the importance of continual surveillance of the use of the substance and scientific advances that affect the safety evaluation process.

A WHO Document (1987) addressed the issue of the use of human studies in safety evaluation and concluded: "Data from controlled human exposure studies are useful if confirming the safety indicated by animal studies after the establishment of ADIs. Such data are also useful in subsequent periodic reviews, and might facilitate a re-evaluation of the safety factors that are applied in calculating ADIs", and further "Prediction and prevention of possible toxic hazards to the community that might arise from the introduction of a chemical into the environment can be made more certain if information from meaningful studies in human subjects is available". [There is] "the need, at a relatively early stage, to obtain information on the absorption, distribution, metabolism and elimination of the chemical in human subjects, since this makes it possible to compare this information with that obtained in various animal species and to choose the species that are most likely to have a high predictive value for human response". For a discussion of ethical considerations concerning the conduct of human studies, the reader is referred to "Proposed international guidelines for biomedical research involving human subjects", a joint project of the World Health Organization and the Council for International Organizations of Medical Sciences, CIOMS, Geneva, 1982 (available from WHO).

The real issue with respect to human studies was discussed by Paget (1970): "The question is not whether or not human subjects should be used in toxicity experiments but rather whether such chemicals, deemed from animal toxicity studies to be relatively safe, should be released first to controlled, carefully monitored groups of human subjects, instead of being released indiscriminately to large populations with no monitoring and with little or no opportunity to observe adverse effects."

<u>References</u>

Paget, G.E. (1970). The design and interpretation of toxicity tests. Methods Toxicol., 1-10.

WHO (1987). Environmental Health Criteria <u>70</u>.

2.6   ADIs, MRLs AND THE PREDICTION OF DIETARY INTAKES

The ADI is allocated by the WHO Group on the basis of extrapolation from animal toxicity data and of evaluation of human exposure data where available. This value is estimated in order to minimize toxicological risk to consumers. The acceptability of MRL's requires that dietary exposure in humans be determined or estimated.

The 1986 JMPR recommended that FAO and WHO should convene a meeting to finalize the details of an acceptable mechanism for predicting potential consumer exposure to pesticide residues. The present meeting welcomed the information that such a meeting would be held at WHO, October 5-8, 1987.

2.7   MAXIMUM REGISTERED USES OF PESTICIDES AND GOOD AGRICULTURAL PRACTICES

The existence of an effective registration procedure and the official infrastructure to control the availability and use of pesticides results in the identification and approval of maximum registered uses of pesticides. These may qualify the maximum rate of use, the maximum number of applications and the minimum interval between the last application and harvest. Other criteria such as methods of application may also be specified by the registration authority.

These national maximum registered uses thus define the maximum use of a pesticide on a crop and are used in acquiring residue data from supervised trials for estimating the resulting maximum residue levels. These levels are usually converted into legal national maximum residue limits (MRLs). On a national basis these MRLs can be (and are) used to monitor/enforce compliance with registered uses. These registered uses are regarded as being included in "good agricultural practice" in the country in which the pesticide is registered.

The Codex Committee on Pesticide Residues (CCPR) is charged with harmonising national MRLs and currently depends on the JMPR for the evaluation of residues data from trials which include maximum registered uses, for the estimation of maximum residue levels and recommendations for internationally acceptable Codex MRLs. Codex MRLS, since they may be based on or derived from data from a range of national registered uses, clearly cannot always be used to require compliance with each of these registered uses. Codex MRLs apply to food commodities in trade and should only be used to establish (by compliance) the acceptability of the commodity in trade.

Nevertheless there has for many years been a tacit assumption that a Codex MRL is derived from a single identifiable agricultural practice or registered use which should be acceptable to all countries. The term "good agricultural practice" has been used to describe the registered uses of pesticides for many years. In 1967 a definition was adopted by JMPR and formed the basis of the current Codex definition. "Good agricultural practice in the use of pesticides is the officially recommended or authorized usage of pesticides under practical conditions at any stage of production, storage, transport, distribution and processing of food, agricultural commodities and animal feed, bearing in mind the variations in requirements within and between regions, which takes into account the minimum quantities necessary to achieve adequate control applied in a manner so as to leave a residue which is the smallest amount practicable and which is toxicologically acceptable". The "officially recommended or authorized usage of pesticides" is that which complies with the procedures, including formulation, dosage rates, frequency of application and pre-harvest intervals, approved by the national authorities.

The Codex MRL is defined as the maximum concentration for a pesticide residue resulting from the use of a pesticide according to Good Agricultural Practice (GAP) that is recommended by the Codex Alimentarius Commission to be legally permitted or recognized as acceptable in or on a food, agricultural commodity, or animal feed. The concentration is expressed in milligrams of pesticide residue per kilogram of the commodity.

The assumption that the maximum registered use in any one situation should always be regarded as "good" practice has often caused problems at CCPR meetings in discussions on MRLs and consumer safety. Codex MRLs are estimated from trials data which include many maximum registered uses. Such limits may be higher than they need be if, as is often claimed, a more rational use of pesticides, advocated as an alternative to the maximum registered use, would result in lower residues.

The meeting recognized that pesticide registration sets criteria for the maximum permitted uses of a pesticide and normally takes into account, at the time of registration, information on efficacy, safety in use, environmental effects and residues in the crop. Good agricultural practice cannot include practices that do not fall within those defined in the conditions of registration.

However, the existing Codex guidelines contain criteria which when considered as part of the overall concept of good agricultural practice can result in a desirable reduction in the use of pesticides.

In deciding on a control practice in any situation a user of pesticides will need to make a choice of:

- active ingredient(s) and registered formulation;
- dosage (and if appropriate, volume);
- number of applications;
- timing and frequency of application;
- equipment and method of application.

In making these choices, decisions should be determined above all by the need to provide efficacious control of the whole pest/disease spectrum with the minimum amount of pesticides. This will be influenced and constrained in particular by:

- conditions of registered use (including risks to operators);
- cost;
- local pest spectrum to be controlled;
- crop (cultivar - and in particular its pest resistance and risk of phytotoxicity);
- availability of cultural or biological means of control;
- compatibility between pesticides;
- potential environmental impact;
- identified side-effects (especially effects on beneficial organisms such as bees, predacious mites and bugs);
- the pesticide resistance of the pest populations.

The meeting recognized the role of advisory/extension services in helping the user to adopt "good" practices taking all the above factors into account. It considered that the real exposure of the human population and the environment to pesticides is determined more by the actual agricultural practice in pesticide use than by the limits to pesticide use set by registration, and that advisory services need adequate support in ensuring that actual agricultural practice in use of pesticides should in fact be good.

It was recognized by the meeting that data from the monitoring of food commodities have regularly shown that, when a pesticide has been found, in relatively few cases (5%) have residues exceeding Codex MRLs and only in about 20% of the samples have residues even approached Codex limits. These observations probably reflect:

1. the distribution of residues data from trials used by the JMPR in estimating maximum residues levels;

2. the actual use of the pesticide by farmers or growers which often may well be less than the maximum registered use on which the MRLs were based.

3. the scale of use of the pesticide.

The meeting proposed that it be clearly understood that MRLs were based on data from supervised trials which included "maximum national registered uses". Such national registered uses are obviously authorised and are recognised as good agricultural practice in the country concerned.

The meeting also recognized the increasing concern about the environmental impact of pesticides and referred to the practice of including, where possible, relevant data such as solubility in water and fate in soils in the JMPR monographs. Accumulating data in many parts of the world now identified unacceptable or "bad" agricultural practices and the meeting recommended that a future meeting of the FAO Panel on Residues of Pesticides in Food and the Environment should study and evaluate these data. In this context the meeting referred to an imminent publication from IUPAC in Pure and Applied Chemistry on the subject of "The Potential Contamination of Ground Water with Pesticides".

The "Seminar on Good Agricultural Practices" proposed to be held at the 1988 CCPR should take these comments into account and seek the views of member countries on this clarification of the significance of MRLs.

Such considerations would have implications for future interpretation and definition of the term "Good Agricultural Practice" and "Maximum Residue Limit" bearing in mind the use of residues levels in the toxicological assessments of the acceptability of a pesticide residue. The meeting recommends that the CCPR should consider re-defining the terms "Good Agriculture Practice" and "Maximum Residue Limit".

2.8   PESTICIDES THAT ARE ALSO METABOLITES OF OTHER PESTICIDES

The meeting noted the principles laid down at the 1979 Joint Meeting (FAO Plant Production and Protection Paper 20, Section 2.7) concerning the expression and definition of residues to which limits refer. The principles describe general considerations for the expression of residues as parent compound and the inclusion of metabolic products in the residue definition.

Metabolites of some pesticides are also registered pesticides. Difficulties are encountered in expressing the residues found following the use of a parent pesticide or its metabolite as a pesticide in terms of Codex MRLs. Analysing food commodities in trade for the metabolite provides no information on which compound was used. However, the document (CX/PR 82/8) prepared by the Codex and FAO (JMPR) Secretariats and endorsed by the 14th Session of the CCPR recognised that, for the purpose of consumer protection and facilitation of international trade in food, the origin of the residue and its relationship to registered uses at the national level should assume less importance than the actual nature and quantity of the residue present in the commodity.

The meeting confirmed its general agreement with the approach suggested by the CCPR in dealing with cases where metabolites are used as pesticides (Report of the 14th Session of the CCPR) and agreed that the following general considerations are basic to estimating MRLs in such cases:

(a)   The chemical entities to which Codex MRLs apply need to be clearly defined, taking into consideration practical limitations of existing analytical procedures and of regulatory capabilities.

(b)   Residues should be defined, as far as possible, in accordance with the principles laid down by the JMPR (FAO Plant Production and Protection Paper 20, Section 2.7), giving special consideration to the use of metabolites as pesticides.

(c)   Whenever possible, the parent pesticide and its metabolite(s) used as pesticides should be subject to separate MRLs taking into account available residues data from supervised trials which must include maximum registered uses.

(d)   Where it is not possible to set separate MRLs (e.g. where the parent pesticide is degraded rapidly), the MRLs applying to the pesticides concerned can only be expressed in terms of the metabolite(s). In these situations the Evaluations should clearly indicate which pesticide was used to obtain the data on which the MRLs had been based. The individual MRLs should be complied with irrespective of the source of the residue. The MRLs should be annotated as follows: "MRL based on data from registered uses of __________" (pesticide(s)).

It was recognized that in setting MRLs for these pesticides, each case would have to be judged on its own merits. In this respect the meeting noted that the 1986 Joint Meeting had developed separate estimates of MRLs for dimethoate and omethoate. The estimates of MRLs for dimethoate were based on information indicating that the metabolite omethoate constitutes only a small part of the total residue. MRLs for omethoate were estimated on the basis of already published information on residues resulting both from the direct application of omethoate and from the metabolism of dimethoate residues. These principles have also been applied by the JMPR to other cases.

2.9 CHEMICAL NOMENCLATURE

To avoid inconsistencies in chemical nomenclature the meeting agreed that the instructions to Authors should include a statement that chemical nomenclature should conform to IUPAC rules and, where these allowed a choice, according to their interpretation by the Technical Committee on Common Names for Pesticides of the International Organization for Standardization (ISO). The responsibility for providing chemical names according to these rules should lie with the FAO Panel.

2.10 INDEX OF EVALUATIONS

A number of compounds have been re-evaluated several times, sometimes toxicologically and sometimes for residues. It was pointed out that the work of referees carrying out re-evaluations would be facilitated if an index to previous evaluations were available which distinguished between those by the WHO Group and by the FAO Panel. The FAO Joint Secretariat undertook to make such an index available, and one is included as Annex II to this Report. This will be updated regularly.

2.11 REPORT OF THE 19TH SESSION OF THE CODEX COMMITTEE ON PESTICIDE RESIDUES (ALINORM 87/24A)

The meeting noted the report of the 19th Session of the CCPR, in particular items drawn to the attention of the JMPR. Several of these were considered at the meeting and are dealt with in the appropriate sections of this report, as follows.

1.   General items:

   - Consideration of recommendations of the Codex Committee on General Principles (paras 235, 284); see 2.7
   - Consideration of a draft paper on pesticide metabolites which are themselves pesticides (para 296); see 2.8

2.   Questions on items in the new Codex Commodity Classification:

   - <u>Fruits, Vegetables</u>: consideration of replacement by narrower groups (para. 31)

   - <u>Meat</u>: consideration of harmonization of MRLs for similar pesticides (para. 33)

- <u>Commodities with indefinite descriptions</u>: clarification (para. 34)

    With certain exceptions among the compounds listed below, all these
    questions required an extensive review of the original data and
    were therefore deferred until 1988.

3.   Questions on the following individual compounds:

    - aldicarb (paras. 117-8). The matter could not be considered owing
      to lack of data.
    - captan (paras. 21, 57); see 4.4
    - carbendazim (paras. 84-5); see 4.5
    - cypermethrin (paras. 119). The matter could not be considered
      owing to lack of data.
    - deltamethrin (paras. 36, 158); see 4.12
    - dichlorvos (para. 63); see 4.
    - dimethoate (paras. 64, 66, 68-9); see 4.14
    - disulfoton (para. 49); see 4.
    - etrimfos (paras. 133-4, 137); see 4.17
    - fenvalerate (para. 125); see 4.20
    - flucythrinate (paras. 185-6, 190); see 4.21
    - folpet (para. 21); see 4.4
    - glyphosate (paras. 199-200); see 4.23
    - heptachlor (para. 73); see 4.24
    - malathion (para. 37). As the necessary original data were not
      available, the matter was deferred until 1988.
    - mecarbam (para. 138); see 4.25
    - metalaxyl (paras. 159, 162-3); see 4.26
    - methiocarb (paras. 147, 149, 151); see 4.27
    - methomyl (paras. 101-3); see 4.28
    - phorate (para. 116). The matter could not be considered owing to
      lack of data.
    - phoxim (para. 168); see 4.34
    - propamocarb (para. 182); see 4.36
    - thiodicarb (paras. 194-5); see 4.40
    - triadimefon (para. 154); see 4.42
    - trichlorfon (para. 38); see 4.43
    The CCPR had noted (para 131) that data on GAP for permethrin in
    tomatoes had been made available to the JMPR, but the data were
    insufficient for evaluation.

3.   <u>SPECIFIC PROBLEMS</u>

3.1   ANALYSIS OF PESTICIDES WHICH YIELD A COMMON ANALYTE

    The residues from the use of vinclozolin are defined by the CCPR as "the
sum of vinclozolin and all metabolites containing 3,5-dichloroaniline,
expressed as vinclozolin". The method of analysis used in developing the data
from which the maximum residues levels were estimated was based on the
hydrolysis of all residues and measurement of the analyte,
3,5-dichloroaniline.

    Residues of at least two other pesticides, iprodione and procymidone,
which also contain the common moiety 3,5-dichloroaniline, would also be
included, if present, in such a "total" method of analysis.

Thus, it is difficult to interpret enforcement or monitoring data using this method of analysis. However, since procymidone and iprodione residues are defined by the CCPR in terms of the parent compound these do not present a problem in themselves. Vinclozolin is metabolised relatively quickly following some uses but can be found as the parent compound. A separate method of analysis for vinclozolin exists.

The meeting confirmed that, as a general principle, MRLs should be defined as the parent compound. Analytical methods based on the determination of an analyte that could be common to several pesticides should be avoided in regulatory situations.

The use of the total 3,5-dichloroaniline method on its own cannot be used to show that the MRL for vinclozolin has been exceeded. Therefore if the "total" method indicates a calculated residue of vinclozolin which exceeds the MRL, it is essential to carry out analyses for the parent compounds concerned, namely, iprodione, procymidone and vinclozolin.

The meeting recognized that this problem occurs in other situations where analysis of pesticides yielded a common analyte, e.g. the determination of CS_2 in the analysis of ethylenebisdithiocarbamates, dimethyl dithiocarbamates, propineb and thiram. It was agreed that this problem should be discussed in detail at a future meeting.

3.2 THE TOXICOLOGICAL SIGNIFICANCE OF TRIAZOLYLALANINE

The Committee noted that a number of compounds (including triadimefon, propiconazole, bitertanol) result in residues of triazolylalanine. This compound is not a metabolite in rodent species. The meeting is aware of the existence of toxicological data on this compound, and recommends that these data be obtained and evaluated as soon as possible to assess the toxicological significance of triazolylalanine.

3.3 ORGANOPHOSPHORUS ESTERS: OPTICAL ISOMERS AND DELAYED NEUROTOXICITY

As previously reported (JMPR 1984), the initiation of delayed neurotoxicity by organophosphorus compounds involves a two-step mechanism: the phosphorylation of the target protein, neuropathy target esterase (NTE), and the "aging" of the phosphorylated protein. It was also suggested that phosphorylation of NTE in delayed neurotoxicity testing provides a quantitative assessment of the potential of a given OP to cause this toxic effect. Recent evidence suggests that when racemic mixtures of phosphonates are used in test animals, the optical isomers might show the same phosphonylating ability for NTE but the rates of aging might differ. Consequently only the optical isomer which forms an ageable protein-phosphonyl complex will cause delayed polyneuropathy. This was the case for EPN, an OP no longer in use. Therefore, whenever OPs are mixtures of optical isomers the delayed neurotoxic potential might depend on the chirality.

4.1   ACEPHATE (095)

Toxicology

Acephate was first evaluated by the JMPR in 1976 when an ADI of 0-0.02 mg/kg bw was allocated. It was re-evaluated in 1982, when invalid IBT studies and the consideration of additional data resulted in the ADI being changed to a temporary ADI of 0-0.003 mg/kg bw. At the 1984 JMPR, in view of the inadequacy of some studies, the safety factor used to calculate the TADI was increased and a lower TADI of 0-0.0005 mg/kg bw was allocated. At the 1984 meeting a multigeneration study, with two litters per generation, and a delayed neurotoxicity study were required. The data required in 1984, together with a lifetime carcinogenicity study in mice, were available and were reviewed by the meeting.

A satisfactory multigeneration reproduction study in rats was reported in 1987 with a no-effect level of 50 ppm (2.5 mg/kg bw/day). A satisfactory delayed neurotoxicity study in Leghorn hens was negative at 785 mg/kg bw/day. A mouse life-time carcinogenicity study showed a significant increase in liver-cell tumours at the highest dose in females with a no-effect level of 36 mg/kg bw/day. An apparent increase in splenic haemangioma/haemangiosarcoma was not considered significant when historical control data were considered.

Toxicological Evaluation

Level causing no toxicological effect

    Mouse: 250 ppm in the diet, equal to 36 mg/kg bw/day
    Rat: 5 ppm in the diet, equivalent to 0.25 mg/kg bw/day
    Dog: 30 ppm in the diet, equivalent to 0.75 mg/kg bw/day

Estimate of acceptable daily intake for man

0-0.003 mg/kg bw.

Studies which will provide information valuable in the continued evaluation of the compound

Observations in man.

4.2   BENALAXYL (155)

Toxicology

Benalaxyl was evaluated toxicologically for the first time by the present meeting. It was evaluated by the 1986 JMPR for residues only. As noted at that time, benalaxyl is completely metabolized and does not accumulate in the tissues of rats. Comprehensive short- and long-term dietary administration indicates that the toxicity of benalaxyl is low. Hepatic enlargement occurred in rats fed benalaxyl (> 1000 ppm) in their diet for 13 weeks and in mice fed for 78 weeks (3000 ppm). However, hepatomegaly did not occur in rats fed benalaxyl (1000 ppm) for two years or in dogs fed for one year (800 ppm).

The meeting concluded that although the hepatic enlargement observed in rodents was of questionable toxicological significance, it could be used to establish a no-effect level and estimate an ADI.

<u>Toxicological Evaluation</u>

<u>Level causing no toxicological effect</u>

    Mouse: 500 ppm in the diet, equivalent to 75 mg/kg bw/day
    Rat: 100 ppm in the diet, equal to 5 mg/kg bw/day
    Dog: 200 ppm in the diet, equal to 7 mg/kg bw/day

<u>Estimate of acceptable daily intake for man</u>

    0-0.05 mg/kg bw.

<u>Studies which will provide information valuable in the continued evaluation
of the compound</u>

    Observations in man.

Residue and Analytical Aspects

    As an ADI has been estimated, the guideline levels for benalaxyl were
converted to MRLs at the same levels.

4.3   BITERTANOL (144) ____

Toxicology

    The 1983 JMPR requested data to clarify the metabolic pathway of
bitertanol and additional toxicological data from a one year dog study and
chronic and carcinogenicity studies in rats.

    In rats, bitertanol is extensively metabolized, primarily by ring and
side chain hydroxylation, oxidation to the corresponding carboxylic acid and
also by ether cleavage. Pharmacokinetic studies demonstrate that it is
absorbed more completely at lower (100 mg/kg bw) than at higher (1000 mg/kg
bw) doses. Tissue retention of absorbed material is low and biliary excretion
predominates.

    Data evaluated by the 1983 JMPR show that short-term feeding of
bitertanol to rats produced body weight depression and hepatotoxicity at 600
and 2400 ppm, (equivalent to 60-240 mg/kg bw/day). Body weight depression was
seen in a 90-day study at 300 ppm (30 mg/kg bw/day). On chronic
administration, bitertanol retarded the growth of rats at 500 ppm (25-50 mg/kg
bw/day).

    In view of the reduced absorption of bitertanol at high doses that has
now been demonstrated, and because of the above mentioned toxicity to rats at
relevant doses, the meeting considered that further chronic or carcinogenicity
studies in rats would be unlikely to yield new information of toxicological
significance.

    An additional study confirms that dogs are more susceptible to
bitertanol than rats. Feeding studies of one and two years' duration show the
formation of cataracts, conjunctivitis and mild hepatotoxicity, with a
no-observed-adverse-effect level at 10 ppm.

Toxicological Evaluation

Levels causing no toxicological effect

    Rat: 20 ppm in the diet, equivalent to 1 mg/kg bw/day
    Dog: 10 ppm in the diet, equivalent to 0.25 mg/kg bw/day

Estimate of acceptable daily intake for man

    0-0.003 mg/kg bw

Studies which will provide information valuable in the continued evaluation of the compound:

    Observations in man.

4.4   CAPTAN (007) / FOLPET (041)

Residue and Analytical Aspects

Although not formally referred by the CCPR at its 19th (1987) Session, the issue was raised concerning the MRLs of captan and folpet in view of their current uses.

Both compounds have been withdrawn from use in Australia, Sweden and Finland and their use in the USA, Canada and EC countries is under review. It has been proposed to reduce the MRLs for these compounds in the EC countries.

The meeting recommended that a detailed review of all aspects of the use of captan and folpet be carried out at the 1989 JMPR or as soon as possible.

4.5   CARBENDAZIM (072)

Residue and Analytical Aspects

The 19th Session of the CCPR (1987) requested the JMPR to consider a proposal that the MRLs for thiophanate-methyl and for carbendazim be combined in a single list under carbendazim.

The meeting referred to the discussion in paragraph 2.8 of this report and proposed that a review of all residues data arising from the use of benomyl, carbendazim and thiophanate-methyl be carried out at the 1988 JMPR.

4.6   CHINOMETHIONAT (080)

Toxicology

The 1984 JMPR withdrew the temporary ADI of chinomethionat because the required data, sufficient to carry out a new toxicological evaluation, had not been supplied. The data supplied to this meeting included studies on reproduction, teratogenicity, mutagenicity, carcinogenicity, a short-term study in rats and a 1-year study in dogs.

In reproduction studies, chinomethionat caused decreased litter-size, pup weight and viability at maternally toxic dose levels. No adverse effects were found at 15 ppm. In teratogenicity studies with rats and rabbits, foetotoxicity and maternal toxicity were observed but no teratogenic effects were seen. The semen of dogs was not affected by chinomethionat administration.

Chinomethionat was not mutagenic in a battery of tests.

In a short-term study in the rat, the main effects were reduced body weight gain, reduced food intake, and anaemia. Marginal effects on red cells were observed at 60 ppm. In a special study with rats, changes in thyroid function were not observed. In a one-year study with dogs, growth inhibition, anaemia, an increase in serum alkaline phosphatase and in alanine amino transferase, as well as histopathological changes in liver and spleen were found. In this study the NOAEL was 25 ppm.

In a long-term toxicity/carcinogenicity study in mice, the same effects were found on body weight and haematological parameters. Relative organ weights were increased, especially those of the spleen. The kidneys showed histopathological changes (progressive nephropathy). With the lowest dose level tested (90 ppm) only marginal haematological changes were observed. The tumour incidence was not enhanced.

The meeting considered these data together with the data previously evaluated in 1974 (acute toxicity and chronic toxicity in rats) and 1977 (biotransformation and acute toxicity of chinomethionat and its main metabolite) sufficient to allocate an ADI. The NOAEL for the rat was based on the long-term study evaluated at the 1974 Meeting, showing no adverse effects at 12 ppm, the highest dose level tested.

Toxicological Evaluation
==
==

Level causing no toxicological effect
==
==

    Rat: 12 ppm in the diet, equivalent to 0.6 mg/kg bw/day
    Dog: 25 ppm in the diet, equivalent to 0.6 mg/kg bw/day

Estimate of acceptable daily intake for man
==
==

    0-0.006 mg/kg bw.

Studies which will provide information valuable in the continued evaluation of the compound
==
==

1. Observations in man.
2. Ongoing studies to identify metabolites in rats and plants.

4.7 CHLORDIMEFORM (013)

Toxicology

Although the results of a urinary monitoring programme have been published, the 1985 Joint Meeting's requirement for epidemiological data on workers occupationally exposed to chlordimeform has not been met.

The results of the Californian monitoring programme indicate that occupational exposure to chlordimeform can be significantly reduced, but not eliminated, during its use in cotton growing. The exposed workers excreted 4-chloro-2-toluidine in their urine.

Workers exposed to 4-chloro-2-toluidine, and to other aniline derivatives in lesser degree, showed a standardized incidence rate of bladder cancer 72 times that of unexposed workers, suggesting that chlordimeform itself could be a human carcinogen.

The meeting agreed to withdraw the temporary ADI (See recommendation, Section 5.2).

Residue and Analytical Aspects

As the temporary ADI for chlordimeform was withdrawn by the meeting, the temporary MRLs previously estimated for this compound were withdrawn. They have not been replaced by Guideline Levels.

4.8   CHLOROTHALONIL (081) ____

Toxicology

Several metabolism studies and the oncogenicity study in mice required by the 1985 JMPR and some additional data have been submitted and evaluated.

^{14}C-Chlorothalonil administered orally as a suspension is incompletely (15-35%) and rapidly absorbed by the g.i. tract. Absorption probably occurs mainly through the small intestine and it is proportionately higher after a small than after a large dose. Chlorothalonil is rapidly distributed to the kidney, where it is covalently bound to proteins but not to DNA.

Approximately 8-12% of an oral dose is excreted in the urine, 17-21% in the bile and 50-61% remains unabsorbed in the faeces. Chlorothalonil metabolites are actively secreted by the kidney. Identification by GC-MS of the urinary metabolites indicates that chlorothalonil is conjugated with lutathione. The 4-hydroxy derivative of chlorothalonil is not a metabolite in non-ruminants. The data available suggest saturable absorption and excretion and a change in metabolism at high doses (between 5 and 50 mg/kg body weight) or after successive doses.

In a 90-day study, where the toxicity of chlorothalonil and that of its monoglutathione conjugate were compared, both compounds induced microscopic changes of the renal tubules. In another short-term study the chlorothalonil-dependent renal lesions showed a change in pattern with time.

In a recent tumourigenicity study in mice chlorothalonil, administered via the diet, was not oncogenic to the kidneys but induced a few papillomas and squamous cell carcinomas in the forestomach which, however, were not considered to be indicative of an oncogenic potential for man. Treatment-related non-neoplastic lesions were seen in the kidneys and in the forestomach. Based on these non-neoplastic gastric lesions the NOAEL in this study was 15 ppm, equal to 1.6 mg/kg bw/day. The absence in this study of treatment-related renal tumours in mice given 750 ppm chlorothalonil in the diet is at variance with the results of a previous mouse oncogenicity study.

Chlorothalonil induced chromosomal aberrations in Chinese hamster ovary cells in vitro, in the absence but not in the presence of metabolic activation by rat kidney S9 mix. However, in a previous chromosomal aberration study in vivo and in several other studies in vitro and in vivo the compound was not mutagenic.

The meeting was informed that a tumourigenicity feeding study in rats is being currently conducted and the final report is expected by December 1988. In consideration of this ongoing study, which is being conducted at lower doses than previous rat or mouse studies, and of the new metabolism data, the meeting agreed that a temporary ADI based on toxicity data for chlorothalonil, not the 4-hydroxy derivative, should be extended. Moreover, because of concern for the demonstrated oncogenicity in rodents, a high safety factor was used.

Toxicological Evaluation

Level causing no toxicological effect

Mouse:	15 ppm in the diet, equal to 1.6 mg/kg bw/day
Rat:	60 ppm in the diet, equivalent to 3 mg/kg bw/day
Dog:	60 ppm in the diet, equivalent to 1.5 mg/kg bw/day

Estimate of temporary acceptable daily intake for man

0-0.003 mg/kg bw.

Studies without which the determination of a full ADI is impracticable

To be submitted to WHO by 1989

1. The ongoing oncogenicity study in rats.
2. Further studies on the mechanism of the organ-specific toxicity of chlorothalonil to the kidney in order to confirm the metabolic pathway responsible for nephrotoxicity.

Studies which will provide information valuable in the continued evaluation of the compound.

Observations in man.

4.9   CLOFENTEZINE (156)

Residue and Analytical Aspects

Clofentezine was reviewed for the first time by the 1986 meeting which required some additional data on registered use patterns and confirmatory residue data on some crops. The data provided in response to these requests enabled the meeting to confirm the maximum residue levels for stone fruits and strawberries estimated in 1986; they are no longer temporary. In addition, maximum residue levels on grapes, currants, raspberries, gooseberries and poultry tissues were estimated and the proposals for cattle tissues were re-examined and amended. Previous recommendations for citrus fruit and cucumbers remain temporary because of the lack of confirmatory data.

Further Work or Information

Required

    1.   Residue data arising from supervised trials of clofentezine on cotton according to registered use patterns.

    2.   Confirmatory data on residues of clofentezine on citrus fruits and cucumbers from supervised trials at registered rates of application.

Desirable

Information on residues of clofentezine present in food in commerce and at consumption.

4.10 COUMAPHOS (018)

Toxicology

Coumaphos was evaluated for toxicological effects by the 1968 Joint Meeting and a temporary ADI was allocated. Because several studies requested in 1972, 1975, 1978 and 1980 were not submitted, the 1980 Meeting withdrew the temporary ADI. Sufficient data have to be available to the JMPR before a full toxicological re-evaluation can be made. The data submitted to the present meeting were considered to be neither qualitatively nor quantitatively sufficient to allocate an ADI.

The meeting agreed that all adequate toxicity data available on the compound, including the oncogenicity and neurotoxicity studies known to be in progress, should be submitted before the allocation of an ADI can be reconsidered. A toxicological monograph was not prepared.

4.11 CYFLUTHRIN (157)

Toxicology

Cyfluthrin has not been previously evaluated by the WHO Expert Group, but was evaluated by the FAO Panel in 1986.

Adequate data were submitted to characterize the rates of absorption and elimination of cyfluthrin. Approximately 60-70% of an oral dose is eliminated via the urine, and the remainder via the faeces. No potential for bioaccumulation is known. Cyfluthrin is readily metabolized, and is excreted as a glucuronide or sulphate conjugate.

Special studies on carcinogenicity, mutagenicity, teratogenicity, and reproductive toxicity demonstrated no effects.

Long-term feeding studies on mice, rats and dogs did not reveal any evidence of organ-specific toxicity. The only significant effects noted in these studies were retardation of weight gain and alterations of organ weights secondary to body-weight effects.

<u>Toxicological Evaluation</u>

<u>Levels causing no toxicological effects</u>

    Mouse: 200 ppm, equivalent to 30 mg/kg bw/day
    Rat: 50 ppm, equal to 2 mg/kg bw/day
    Dog: 160 ppm, equal to 5.1 mg/kg bw/day

<u>Estimate of acceptable daily intake for man</u>

    0-0.02 mg/kg bw.

<u>Studies which will provide information valuable in the continued evaluation of the compound</u>

    Observations in man.

4.12 DELTAMETHRIN (135)

Residue and Analytical Aspects

In response to the requirement by the 1986 JMPR for data on residues resulting from recommended uses of deltamethrin on stored cereal grains, extensive data were provided on a variety of stored commodities following treatments with deltamethrin according to registered uses or in large scale trials. The meeting also received information on registered uses in New Zealand, and on the registered uses on hops in several countries.

From the data on residues on stored grains the meeting concluded that:

1.   The maximum recommended dose rate for deltamethrin for the protection of a number of stored commodities is 0.5 mg/kg with piperonyl butoxide (PB) or 1 mg/kg without PB.

2.   Residues on cereal grains and other commodities vary from about 20% to 80% of the treatment dose. There is variation in this range between commodities and from one country to another, probably reflecting the method and efficiency of the application technique. Residues found were never in excess of the stated application rate.

3.   The "lost" deltamethrin is associated with "impurities" such as dust and other plant and mineral debris in the commodity. This is presumably still effective in insect control.

4.   In trials on stored rice in Brazil, France and, particularly, the People's Republic of China when the rice was stored as paddy, the data show that the main part of the residue is on the husk and to a lesser extent in the bran. Husking therefore removes most of the residue and most of the remainder is eliminated by polishing the brown rice. Thus polished rice may contain up to 10-15% of the initial deposit.

5.   Residues in the grain at application show no appreciable loss over 1 year in storage. Variations in residues found probably reflect distribution and sampling problems.

6.   Cooking polished rice by steaming reduces residue levels to close to the limit of determination.

7.   Deltamethrin is not degraded significantly in any of the cooking/processing operations examined for wheat.

Accepting that the maximum dose rate was 1 mg/kg, the 1986 JMPR referred to the suggested MRL of 2 mg/kg for cereal grains intended to cover variation due to uneven distribution that is likely to occur. The new data indicate clearly that the distribution of deltamethrin is undoubtedly uneven but, with "losses" in the bulk which can be identified, the measured levels did not exceed the initial intended dose. It therefore seems unlikely that regulatory sampling would raise a problem that, once identified, could not be resolved by further sampling, and the meeting agreed that an estimated maximum residue level of 1 mg/kg would be suitable for an MRL. Data on the fate of deltamethrin on processing wheat showed that the following maximum residues levels could be estimated in various processing fractions: wheat bran, unprocessed 2 (previously 5); wheat wholemeal 0.5 (previously 1 for wheat flour, wholemeal); wheat flour (white) 0.1 (previously 0.5).

New data on dried hops demonstrated that, at registered rate of use, the present proposal of 5 mg/kg was adequate. As expected from the properties of deltamethrin the major part of the deltamethrin residue is eliminated by the brewing process in the lees, a little is found in the yeast and none in beer. In hop-tea about 12-14% of the deltamethrin residue is in the extract.

The meeting also examined a previous recommendation for "assorted fruits - edible peel, 0.1 mg/kg". It commented that this group definition would include many and varied fruits with widely different residue characteristics and was in any case not recognized in the new classification of commodities. After a re-examination of the data on the two crops originally used in formulating the proposal, the meeting agreed to recommend separate figures for figs and olives.

4.13 DIMETHIPIN (151)

Toxicology

Dimethipin was on the agenda of the present meeting because the 1985 Joint Meeting required further data by 1987. However, the results of the requested studies were made available too late for consideration by the present meeting. The meeting agreed to extend the temporary ADI of 0-0.003 mg/kg bw for one year to permit full evaluation of the requested data.

Residue and Analytical Aspects

Dimethipin was evaluated by the 1985 JMPR; a temporary ADI was estimated and some residue levels were recommended as suitable for use as temporary MRLs. The 1985 JMPR concluded that these limits should remain temporary irrespective of the status of the ADI, pending the receipt of the required information. Information was provided on most of the questions. Data were provided from several countries on the authorised or otherwise recommended uses of dimethipin as a defoliant or desiccant on cotton, potatoes, tomatoes, rice and oil seed crops, including sunflower, shortly before harvest (10-14 days).

Additional data on residues from supervised trials on sunflower treated according to GAP were received from one country, from 5 different sites. Residues of the parent compound at harvest ranged between 0.01 and 0.38 mg/kg.

Extensive additional information was provided on the fate of dimethipin in cotton seedlings and callus cell cultures of cotton and potato. Several metabolites were isolated by HPLC and identified by co-chromatographic procedures. The main metabolite in the plant material, accounting for about 70.4% of the extractable radioactivity from the cotton seedlings, was the parent compound. This was confirmed by mass spectrometry and NMR. The remaining extractable radioactivity included 4-5 metabolites which could be identified by co-chromatography. This provided clear evidence that most of the plant metabolites result from an initial conjugation of dimethipin with the plant constituents glutathione and/or cysteine and subsequent cleavage of the conjugates to yield cyclic and acyclic derivatives of dimethipin.

Comparison of the metabolism in plants and plant cell cultures with that in animals (goat tissues, rat urine) provided evidence that the metabolic pathway of dimethipin in plants is similar to one of the pathways elucidated in animals, as described in the 1985 Evaluations (from a study in goats and the identification of metabolites in the rat urine).

Feeding trials were carried out in which goats received normal feed ration supplemented with 1 and 10 mg/kg ^{14}C-dimethipin in the feed. The feeding was continued until a plateau level was reached after 17 days in the milk. At the end of the experiment, after 18 days of dosing, about 97% of the total administered radioactivity was excreted via the urine and faeces. On the 18th day residues were not detactable in muscle and were low in milk, liver and kidney. The feeding levels were high compared with a so-called "worst case" feeding situation in which every component of the diet contain residues at the level of the TMRL proposed by the 1985 JMPR. The residue levels which can be expected even in such a "worst case" situation will therefore be much lower than those found in the experiment and thus if animals are exposed to a normal feeding regime in practice the levels in meat, milk and edible offal will not exceed the limit of determination.

Laying hens were fed a normal poultry diet fortified with ^{14}C-dimethipin at dosage rates of 1, 6 and 30 mg/kg in the feed for 30 days. Half the animals were slaughtered after 30 days and the remainder after a withdrawl period of 11 days. By day 30 more than 95% of the total administered radiocarbon had been excreted via the faeces. Residues in muscle and fat were not detectable at the termination of dosing in the 1 mg/kg group or in the fat of the 6 mg/kg group. The levels found in tissues corresponded roughly with the dose levels. The levels in the tissues of all animals analysed after 11 days of withdrawal were all considerably lower. None of the tissues from the 1 mg/kg group, except the blood, contained any measurable residues. The animals fed with 6 and 30 mg/kg ^{14}C-dimethipin had measurable residues in all tissues except the fat. In eggs the residue levels (as radiocarbon) increased slowly over about 10 days and remained fairly constant for the remainder of the dosing period. Except in the 30 mg/kg group, the levels in eggs decreased to below the limit of determination during the 11 days withdrawal. By extrapolation from the data obtained it can be concluded that residues in animals fed according to a hypothetical "worst case" diet in which all or nearly all components contain residues at the level of the MRL proposed by the 1985 JMPR would not exceed the limit of determination.

The new information on residues and metabolism in animals satisfied the requirements of the 1985 JMPR in this area.

The degradation of dimethipin in soil under aerobic conditions follows a different pattern compared with the metabolism in plants and animals. After isomerisation of dimethipin, degradation proceeds via the carboxylic derivative of dimethipin to 1,4-dithiane 1,1,4,4-tetraoxide and further ring fragmentation into more polar unidentified breakdown products.

On the basis of the new information on residues from supervised trials on sunflower, the information available to the 1985 JMPR on sunflower seed oil and the information on residues in products of animal origin, the meeting estimated a revised maximum residue level for sunflower seed and additional levels for sunflower seed oil and products of animal origin. The estimate for potato was revised in the light of the current limit of determination. These estimates are recommended for use as TMRLs. All limits are now temporary only because the ADI remains temporary.

4.14 DIMETHOATE (027)

Toxicology

The 1984 JMPR replaced the ADI of 0.02 mg/kg bw by a temporary ADI of 0.002 mg/kg bw because the data on which the former ADI was based were considered inadequate. The meeting was aware that new toxicological studies were under way. These studies were reviewed by the present meeting.

In teratogenicity studies with rats and rabbits maternal toxicity and foetotoxicity were observed, but no teratogenic effects.

In 1984 the Joint Meeting concluded that dimethoate was mutagenic in a limited number of in vitro and in vivo studies reported in the literature. Since then, additional negative mutagenicity data have become available. It was therefore concluded that dimethoate is mutagenic in bacterial tests, but not in mammalian cells or in in vivo tests.

In long-term toxicity/carcinogenicity studies with mice and rats no indication of carcinogenicity was found. In the mouse study the lowest dose of 25 ppm produced decreased body weight and decreased cholinesterase activities in erythrocytes as well as slight extramedullary haematopoiesis in the spleen.

Two years' administration of dimethoate to rats also revealed a decrease in body weight, decreased cholinesterase activities (in erythrocytes and brain) and slight anaemia. No effects were observed at 1 ppm (equivalent to 0.05 mg/kg bw/day).

The required dog study was not available.

The 1984 evaluation of human data was based on the understanding that plasma CHE activity was measured. Re-examination of the data showed that they were in fact based on erythrocyte CHE and that a NOAEL was present. The present meeting therefore decided that an ADI could be established on the basis of these data.

<u>Toxicological Evaluation</u>

<u>Level causing no toxicological effect</u>

    Rat: 1 ppm in the diet, equivalent to 0.05 mg/kg bw/day
    Man: 0.2 mg/kg bw/day

<u>Estimate of acceptable daily intake for man</u>

    0-0.01 mg/kg bw.

<u>Studies which will provide information valuable in the continued evaluation of the compound</u>

    1.   Observations in man.

    2.   A one-year dog study.

    3.   An up-to-date reproduction study in rats.

Residue and Analytical Aspects

In response to the request by the 19th Session of the CCPR for additional information on GAP from member countries, data were submitted for France, Greece and the USA. Uses on commodities other than those with MRLs are shown. It would be desirable to review residues data on selected major crops for the purpose of estimating maximum residue levels. Some uses are also indicated in the Codex system which are not included in the reported GAP of the three countries. Data from more countries appear to be necessary in order to assess general GAP in the use of dimethoate.

In the absence of additional information on residues from supervised trials, the meeting decided not to change the existing MRL for Witloof chicory of 0.5 mg/kg.

<u>Further Work or Information</u>

<u>Desirable</u>

    1.   Residues from supervised trials with nectarines.

    2.   Data on residues from supervised trials on major crops.

    3.   National use patterns.

4.15 ETHOPROPHOS (149)

Toxicology

Toxicology data for ethoprophos were evaluated by the 1983 JMPR. An ADI was not allocated by that meeting because the available data were inadequate. Supplementary data were submitted to the present meeting.

An acute delayed neurotoxicity study in the hen did not reveal any clear evidence of a neurotoxic response. However, because of the high mortality in this study, and equivocal findings in some of the treated birds, data on the effect of this compound on neuropathy target esterase (NTE) activity in the hen would be useful.

A two-year feeding study in the mouse did not reveal any evidence of a
carcinogenic effect of ethoprophos in this species. A NOAEL of 0.2 ppm (equal
to 0.035 mg/kg bw/day) was established for erythrocyte cholinesterase
inhibition.

A two-year feeding study in rats produced equivocal evidence of a
carcinogenic response in the thyroid of high-dose males, but this effect was
restricted to the high-dose group. No other treatment-related lesions were
identified. The overall NOAEL for this study was determined to be 1 ppm
(equal to 0.05 mg/kg bw/day), based on the inhibition of erythrocyte
cholinesterase activity at higher doses.

A 52-week oral dose study in dogs provided evidence of a hepatotoxic
response in mid- and high-dose males and females as indicated by alterations
in liver histology accompanied by disturbances in related serum chemistry
values. The NOAEL for this study was determined to be 0.025 mg/kg bw/day,
based on inhibition of erythrocyte cholinesterase activity and evidence of
hepatotoxicity at higher doses.

Toxicological Evaluation

Level causing no toxicological effect

Mouse:
0.2 ppm in the diet, equal to 0.035 mg/kg bw/day
Rat: 1 ppm in the diet, equal to 0.05 mg/kg bw/day
Dog: 0.025 mg/kg bw/day

Estimate of acceptable daily intake for man

0-0.0003 mg/kg bw.

Studies which will provide information valuable in the continued evaluation of the compound

1.   Observations in man.

2.   Data on the effect of ethoprophos on neuropathy target esterase
     (NTE) activity in the hen.

Residue and Analytical Aspects

As an ADI has been estimated, the guideline levels for ethoprophos were
converted to MRLs at the same levels.

4.16   ETHYLENETHIOUREA (ETU) (108)

Residue and Analytical Aspects

The CCPR at its 19th (1987) Session concluded that as banana pulp was
not a commodity in international trade, MRLs for banana pulp should be
deleted. Where possible, Codex MRLs should be established for whole bananas.
It was understood that the residue levels occurring in the pulp should be
presented as information in JMPR monographs.

The 1974 JMPR recommended a TMRL for ETU in banana pulp (associated with the limit for mancozeb), based on data on residues from supervised trials with mancozeb and the ETU metabolite occurring at harvest. No data were available on residues in whole banana or in the peel.

The meeting was aware that the data on residues of ETU were from one country only and obtained by a method of analysis which would not meet current quality standards. The data may therefore not be reliable.

Because no data were available on residues of ETU in whole bananas or in banana peel and in view of the doubtful reliability of the method of analysis used, the meeting concluded that the MRL for banana pulp should be withdrawn.

4.17 ETRIMFOS (123)

Residue and analytical aspects

The CCPR at its 19th Session expressed the view that the limit of determination proposed by the JMPR for several commodities, was too low to be easily attainable in regulatory laboratories.

The meeting recognized this difficulty and concluded that the residue should preferably be based on the parent compound only. To enable this to be done a review of all data would be necessary and therefore the meeting postponed the complete review to a future meeting, for which the information on GAP desired at the 1986 JMPR would also be included.

The request to change the limits for kale, onions and potatoes was also deferred until the future meeting in the light of the preferred change of the residue definition.

4.18 FENAMIPHOS (085)

Toxicology

The oncogenicity study in rats, the full report of the teratology study in rats and the new teratology study in rabbits required by the 1985 JMPR have been submitted and evaluated.

Fenamiphos was not oncogenic to Fischer 344 rats in a combined chronic toxicity/oncogenicity study. Erythrocyte cholinesterase activity was inhibited in both sexes at all dose levels tested. No biologically significant cholinesterase inhibition was seen, however, in a 90-day rat study at doses up to and including 0.9 ppm, equal to 0.07-0.08 mg/kg bw/day.

Despite some inconsistencies noted in the full report of the rat embryotoxicity/teratogenicity study required by the 1985 JMPR, this study was considered to be adequate. The compound was not teratogenic to Long Evans FB 30 rats, but it was toxic to the dams at the highest dose of 3 mg/kg bw/day.

In the required new teratology study (JMPR 1985), which was performed in Chinchilla rabbits, fenamiphos was toxic to the mothers but it was not embryotoxic or teratogenic. This was in contrast to the results of a previous teratology study in New Zealand rabbits, where fenamiphos administered at similar or lower doses was found to be both foetotoxic and teratogenic.

A re-evaluation and comparison of data from all these studies indicate that fenamiphos is toxic to female rats and rabbits. However, the concern of the 1985 Meeting about the foetotoxicity and teratogenicity observed in New Zealand rabbits was partially alleviated by the results of two new studies. A 3-generation reproduction study previously evaluated by the JMPR was negative.

On the basis of the results of the oncogenicity study in rats and on the species- and strain-dependent results of the foetotoxicity/teratogenicity studies, an ADI was estimated.

Toxicological evaluation

Level causing no toxicological effect

    Rat: 0.9 ppm in the diet, equal to 0.07 mg/kg bw/day in males, and 0.08 mg/kg bw/day in females

    Dog: 2 ppm in the diet, equivalent to 0.05 mg/kg bw/day

Estimate of acceptable daily intake for man

0-0.0005 mg/kg bw

Studies which will provide information valuable in the continued evaluation of the compound

Observations in man.

4.19 FENITROTHION (037)

Residue and Analytical Aspects

Fenitrothion has been reviewed a number of times from 1969 to 1984 and an ADI has been estimated. MRLs have been estimated for cereal grain and wheat flour (white).

At the 19th (1987) Session of the CCPR it was proposed that the MRL for wheat flour be reviewed. Fenitrothion is used as a grain protectant and preliminary information was received by the meeting on a wheat storage trial in Australia in 1987.

The Australian Wheat Board arranged the treatment of 612 tonnes of wheat with "d-phenothrin" (nominal 1 mg/kg) and fenitrothion (nominal 12 mg/kg). At approximately 1.4 and 6 months after treatment 1 tonne samples were milled through a pilot mill (capacity 660 kg/hour) to produce wheat fractions for analysis in two laboratories, one in Australia and the other in Japan. The trial has not yet been completed.

Fenitrothion residues in the wheat were 7.4 and 7.7 mg/kg after 4.6 and 17.4 weeks' storage respectively. White flour produced from this wheat averaged 1.7 mg/kg fenitrothion (range 1.2 - 2.1 mg/kg).

The Codex MRL for wheat flour (white) is 1 mg/kg with a proposed amendment to 3 mg/kg (JMPR 1979). Residues in flour in the trial were more in keeping with the proposed amendment.

4.20 FENVALERATE (119)

Residue and Analytical Aspects

The 19th (1987) Session of the CCPR requested the JMPR to re-examine the proposed MRL for edible offal (mammalian) of 0.02 mg/kg.

The meeting noted that the subject of animal transfer studies and the estimation of MRLs in foods of animal origin was reviewed in some detail by the 1986 JMPR and, although residues of fenvalerate in edible offal were not referred to specifically in the 1986 Report, the 1986 Evaluation confirmed that the recommendations already made for MRLs in edible offal were appropriate. The meeting re-confirmed this view.

The data on Brussels sprouts, promised at the 19th Session of the CCPR, were not received by the meeting.

4.21 FLUCYTHRINATE (152)

Residue and Analytical Aspects

Flucythrinate was evaluated by the 1985 JMPR. An ADI was allocated and several maximum residue levels suitable for use as MRLs estimated. The 1985 JMPR required additional information on the fate of flucythrinate residues during the processing of oil seeds into crude and refined vegetable oils. Also additional information was desired on residues from supervised trials on main crops on which flucythrinate is used according to GAP, including information relating to glasshouse lettuce, and residues found in meat and other products of animal origin after feeding treated animal feed crops or other crops of which wastes are used as animal feed. The CCPR at its 19th 1987 Session noted the desirability of obtaining residue data on maize fodder on a dry weight basis.

Extensive data were obtained on current GAP for flucythrinate in the USA on pome fruit, head cabbage, lettuce, maize (including "sweet corn") and cotton, and on the fate of residues during the processing of cotton seed into cotton seed oil and other products.

The additional data confirm in general the previously recommended limits for pome fruit and cotton seed. Since no data were received on residues from supervised trials on glasshouse lettuce no limits could be proposed for head lettuce, grown either in the open or under glass.

In the absence of additional data on residues in products of animal origin the MRLs for these commodities proposed by the 1985 JMPR remain temporary.

The meeting noted that the limit of 0.5 mg/kg for beans (dry), proposed by the 1985 JMPR, was a typographical error. The intended estimate was evidently 0.05* mg/kg.

The limit of determination by the analytical method of choice, namely GLC with a 63 Ni electroncapture detector, was validated for apples, head cabbage, lettuce, maize (including forage and fodder), cotton seed and cotton seed oil. In all the commodities mentioned, the validated limit of determination was 0.05 mg/kg.

On the basis of the combined 1985 and 1987 data together with the updated information on GAP the meeting estimated a revised maximum residue level for head cabbages and additional levels for maize, sweet corn (kernels), maize fodder and cotton seed oil, which are suitable for use as MRLs.

<u>Further Work or Information</u>

<u>Desirable</u>

1.   Additional information on residues from supervised trials on main crops on which flucythrinate is used e.g. various citrus species, leafy vegetables such as lettuce, including glasshouse lettuce, fruiting vegetables, including glasshouse crops, legume vegetables (pods and seeds).

2.   Information on actual residues found in meat, fat, milk and eggs after feeding treated animal feed crops or other crops of which wastes are used as animal feeds, e.g. straws of cereal grains, vines or straws of legume vegetables etc.

4.22 FOLPET (041)

Residue and Analytical Aspects

See 4.4, captan/folpet.

The meeting recommended that a detailed review of all aspects of the use of folpet and captan be carried out at the 1989 JMPR or as soon as possible.

4.23 GLYPHOSATE (158)

Residue and Analytical Aspects

Glyphosate was evaluated by the 1986 JMPR, an ADI was estimated and MRLs were recommended for some cereal grains, vegetables, oilseeds and animal products. The limits were for glyphosate only.

Information was received in response to the requirements of the 1986 JMPR for a robust analytical residue method suitable for a wide range of substrates, and for residue data on milled products and processed cereal commodities.

Residues in treated cereals were consistent with levels from previously evaluated trials.

Residue levels in white flour were approximately 10-20% of those in wheat, while the bran levels were 2-4 times as high as those in the wheat. The highest bran residue was 26 mg/kg arising from wheat containing 6.8 mg/kg glyphosate.

Glyphosate residues were not lost during baking, but residue levels decreased on making bread from flour because of dilution. The metabolite AMPA (aminomethylphosphonic acid) was not detected in flour or bread. In bran AMPA was present at approximately 2% of the glyphosate level.

In answer to a CCPR request for reconsideration of the 20 mg/kg MRL on wheat, the relation between the present use pattern and the trials data was re-examined. The label instruction for pre-harvest use on cereals now specifies that the least mature cereal grain must be under 30% moisture content when glyphosate is applied. On the revised use pattern the meeting was able to exclude trial data on wheat in the 10-20 mg/kg range and estimate a revised maximum residue level for wheat.

Glyphosate residue levels in malt and beer derived from field-treated barley were respectively 25% and 4% of the original level in the barley. Some glyphosate is lost during washing, but most of the decrease can be attributed to dilution. AMPA was not detected in malt or beer. Spirits prepared from malt containing glyphosate residues had no detectable residues (< 0.05 mg/kg).

Residue levels in groats (processed oats) were about 50% of the levels in the pre-harvest-treated oats. The ratio of AMPA to glyphosate levels was substantially unchanged during processing.

In the USA the use of glyphosate as a directed application to weeds in soya bean crops has been developed, at rates of 0.2-4.2 kg ai/ha. It is not the intention to apply the herbicide to the crop but some contamination occurs from splatter off the weeds or direct physical contact of weeds and crop.

In soya bean grain from treated fields, glyphosate residue levels up to 3.3 mg/kg were observed. Unlike the situation with other crops, residues of AMPA were of the same order as those of glyphosate. Metabolism in soya bean may be atypical and should be investigated further. In soya bean hay and forage the ratio of AMPA to glyphosate was similar to that in other crops.

Residues up to 11.7 mg/kg were observed in soya bean hay. Residues were not detectable (< 0.05 mg/kg) in oil prepared from soya bean grain samples containing 1.2-3.9 mg/kg glyphosate and 0.6-1.2 mg/kg AMPA.

In New Zealand, glyphosate is used in kiwi fruit orchards as a directed spray to control weeds. The spray is not allowed to contact foliage or green stems of the crop. Residues were not detected in kiwi fruit (< 0.05 mg/kg).

A residue analytical method suitable for a range of substrates has been collaboratively tested. The method depends on chelation ion-exchange and anion-exchange for clean-up, followed by HPLC separation and fluorescence detection after post-column derivatization.

Five laboratories were involved in the collaborative study on five sample types at concentrations of 0.05-5 mg/kg for both glyphosate and AMPA. All analysts achieved overall average recoveries greater than 70%.

This method will be valuable to regulatory laboratories monitoring a variety of sample types.

4.24 HEPTACHLOR (043)

Residue and Analytical Aspects

The CCPR at its 19th (1987) Session discussed a proposal to delete the words "(in the edible portion)" from the Codex limit for heptachlor in pineapples, and implicitly sought the views of the JMPR on this proposal.

The meeting noted that all the limits for heptachlor were ERLs and, apart from those in fatty materials, at low levels. The compound is no longer used, residues in pineapples would occur only as a result of uptake from the soil, and residue levels in the peel would probably not differ from those in the flesh. The meeting therefore concluded that the words "(in the edible portion)" could be deleted.

4.25 MECARBAM (124)

Residue and Analytical Aspects

The CCPR at its 19th Session requested the JMPR to reconsider the proposed limits for cattle meat, edible offal of cattle, and cattle milk. The proposed limit for milk was 0.01 mg/kg and those for cattle meat and edible offal of cattle were at the limit of determination of 0.005 mg/kg. The CCPR was of the opinion that the proposed limits were too low to be practicable for regulatory analysis. A limit of 0.02 mg/kg was proposed.

The meeting reviewed the data on the method of analysis and the limit of determination for the commodities mentioned and agreed to increase the limits for cattle meat and edible offal of cattle from 0.005 mg/kg to 0.01 mg/kg but to retain the limit for milk at 0.01 mg/kg.

4.26 METALAXYL (138)

Residue and Analytical Aspects

The meeting took note of the strong objections that had been expressed by several delegations at the 19th Session of the CCPR to the change in the definition of the residue that was recommended by the 1986 JMPR. This new recommendation was that the definition should include the "total" residues of the parent compound together with certain of its metabolites instead of referring to the parent compound only. This change was made at that meeting because the preferred method of analysis determined such "total" residues after hydrolysis to 2,6-dimethylaniline and much of the residue data submitted had been obtained using this method.

Having reconsidered its position on the matter, the meeting agreed that there was merit in the views of the CCPR and decided that it would now express the residues of metalaxyl in terms of the parent compound alone. This reversal of the residue definition, returning to that used at the 1982, 1984 and 1985 meetings, had little or no effect on the maximum residue levels estimated for most crops.

In the light of this decision, the meeting considered the additional data that had been submitted this year and also reviewed again the data submitted previously which are detailed in the Evaluations of the meetings held in 1982, 1985, and 1986. As a result of these reviews, a complete list of maximum residues levels has been compiled; this includes some crops for which new data had become available and for which no recommendations had previously been made. It replaces all previous lists (see Annex I).

Unfortunately, apart from limited data on lettuce and spinach, no additional data on residues in leafy vegetables were submitted and therefore the meeting was not able to develop any new proposals for such crops. All recommendations for residue levels quoted for the brassica vegetables are therefore based only on a review of the data published previously; all such levels are unchanged from those estimated in earlier years.

Further Work or Information

Desirable

1.   Data on residues of metalaxyl on brassica and other leafy crops following applications at recommended treatment rates, using a method of analysis that determines the parent compound.

2.   Confirmatory data on residues of metalaxyl in cacao beans, peppers and avocados.

3.   Information on residues of metalaxyl in food in commerce or at consumption.

4.27 METHIOCARB (132)

Toxicology

Additional data on the method used to measure cholinesterase activity in the long-term dog study were identified as being desirable by the 1983 JMPR. Data submitted to the present meeting did not fully address the concerns of the previous JMPR, but in view of the large amount of data available for this compound the meeting agreed to maintain the ADI at its present level.

Toxicological Evaluation

Levels causing no toxicological effect

    Rat: 25 ppm in the diet, equivalent to 1.3 mg/kg bw/day
    Dog: 5 ppm in the diet, equivalent to 0.13 mg/kg bw/day

Estimate of acceptable daily intake for man

    0-0.001 mg/kg bw

Studies which will provide information valuable in the continued evaluation of the compound:

    Observations in man.

Residue and Analytical Aspects

The meeting reviewed the current important registered uses of methiocarb and the relevant residues data. It recommended the confirmation of a number of MRLs to reflect these uses and the withdrawal of certain others for commodities on which there no longer appeared to be any significant use.

Further Work or Information

Required

Residues data on artichokes and hazel-nuts from residues trials which include registered uses to enable a future meeting to estimate maximum residues levels.

4.28 METHOMYL (094)

Residue and Analytical Aspects

A review of the data summarized in the 1975 monographs together with new residues data on cereal grains and hops enabled the meeting to estimate an increased maximum residue level of 2 mg/kg of methomyl in hops with an interval of 10 days (the previous temporary MRL was 1 mg/kg with an interval of 14 days).

The meeting also agreed that an estimated maximum residue level of 0.2 mg/kg was appropriate to barley, oats and wheat compared with the previous levels of barley 0.2, oats 0.1 and wheat 0.1 mg/kg. The new figure is based on an interval of 7 days between last application and harvesting compared with a 14-day interval for the previous estimates.

4.29 METHOPRENE (147)

Toxicology

The 1984 JMPR required a 6-month feeding study in dogs, adequate teratology studies (dosing in the organogenetic period), and a 2-generation (2 litters/generation) reproduction study in rats.

The meeting reviewed the data, in particular the metabolic and kinetic data, which indicate that the compound is completely degraded. It seemed unlikely that the compound would reach the conceptus in teratology and reproduction studies. Long-term studies in dogs would not give further information.

Toxicological Evaluation

Level causing no toxicological effect

Rat: 500 ppm in the diet, equivalent to 25 mg/kg bw/day
Dog: 500 ppm in the diet, equivalent to 12.5 mg/kg bw/day

Estimate of acceptable daily intake for man

0-0.1 mg/kg bw.

Studies which will provide information valuable in the continued evaluation of the compound

Observations in man.

4.30 OMETHOATE (055)

Residue and Analytical Aspects

Data from supervised trials on hops in the Federal Republic of Germany could not be evaluated because unexplained high residues appeared to be present in control samples. A detailed trial on Valencia oranges in Portugal provided data that supported the existing MRL of 2 mg/kg for citrus fruit. Detailed monitoring studies on a range of crops in Finland confirmed that omethoate is rarely found in food commodities imported into Finland.

4.31 PERMETHRIN (120)

Toxicology

As noted by the 1982 JMPR, the acute toxicity of permethrin (25:75 cis-, trans-) is less than that of permethrin (40:60). On short-term administration to rats and mice, permethrin (25:75) was of similar toxicity to permethrin (40:60). Permethrin (25:75) is not mutagenic in short term tests or in a dominant lethal assay and this isomeric mixture is not carcinogenic in mice or rats. It is also not teratogenic in, and does not adversely affect the reproduction of, rats. Chronic feeding of permethrin (25:75) to rats for 2 years increased the liver weights of male rats, as did permethrin (40:60).

The toxicological profile of permethrin (25:75) thus resembles that of permethrin (40:60), although it is less acutely toxic. Accordingly, the meeting agreed to extend the previously established ADI for permethrin (40:60) to include permethrin (25:75).

Toxicological Evaluation

Level causing no toxicological effect (for the nominal 40% cis-, 60% trans- and 25% cis-, 75% trans- materials)

Rat: 100 ppm in the diet, equivalent to 5.0 mg/kg bw/day

Estimate of acceptable daily intake for man

0-0.05 mg/kg bw.

Studies which will provide information valuable in the continued evaluation of this compound

Observations in man.

4.32 PHENOTHRIN (127)

Residue and Analytical Aspects

Phenothrin has been evaluated by the JMPR several times since 1979. Temporary MRLs have been recommended for cereal grains (5 mg/kg) and wheat bran (15 mg/kg). Further residue data on stored products and improved methods of residue analysis were requested. The 1986 CCPR agreed to discuss the MRLs for wheat bran and cereal grains when data on wheat flour had been considered by the JMPR. Information on a milling and baking trial and methods of residue analysis were made available to the meeting.

In Australia, "d-phenothrin" (cis/trans ratio = 2/8) is recommended as a grain protectant in combination with piperonyl butoxide and an OP. For grain to be stored longer than 3 months the recommended treatment rate is 1 mg/kg.

In 1987 the Australian Wheat Board arranged the treatment of 612 tonnes of wheat with "d-phenothrin", piperonyl butoxide and fenitrothion. At approximately 1, 4 and 6 months after treatment, 1-tonne samples were milled through a pilot mill (capacity 660 kg/h) to produce wheat fractions for analysis in two laboratories, one in Australia and the other in Japan. Analytical results from the two laboratories using different methods were in good agreement.

There was no evidence of residue decline during storage in the silo. Unfortunately, the actual treatment rate (approximately 0.55 mg/kg) was much lower than the nominal rate (1 mg/kg). Residues in the wheat fractions would be expected to be proportionally higher than those found in the trials. Residue levels in the fractions were not influenced by the different milling extraction rates tested (75 and 80%).

Levels in white flour ranged from 0.15 to 0.22 mg/kg, which were approximately 1/3 of the residues in the wheat. In bran, phenothrin residues ranged from 1.9 to 2.7 mg/kg which were an average of 3.8 times those in the wheat. The decline in phenothrin residues during baking was almost all accounted for by dilution. Phenothrin residues in the germ were on average twice the levels in the grain. The range of residue concentrations in germ was 0.38 to 1.6 mg/kg.

A laboratory study of phenothrin residues on wheat in jar trials under controlled conditions of temperature and moisture showed that phenothrin was quite persistent for most practical storage times and conditions, e.g. a half-life of 72 weeks was found for conditions of 25°C and 12% moisture.

Descriptions of three analytical residue methods suitable for trials or regulatory and survey work were made available to the meeting. Each method determines phenothrin as the parent molecule. In the first the sample was extracted with methanol. Clean-up was effected by solvent partition and an alumina adsorption column. Phenothrin was then determined by HPLC on a reversed phase system (C18 column with a methanol/water eluent) using a UV detector at 236 nm. The second method used acetone/water for sample extraction, then solvent partition and an adsorption column for clean-up. Final analysis was by GLC with mass spectrometer monitoring of the fragment m/e 183. Hexane was used for extraction in the third method. The clean-up was minimal, consisting of filtration through a 0.45 μm filter. Phenothrin residues were then determined by HPLC on a silica column eluted with 0.25% tetrahydrofurane in hexane with UV detection at 230 nm.

Further Work or Information

Required

A milling and baking trial with "d-phenothrin" applied to stored wheat at the recommended rate of 1 mg/kg.

4.33 PHOSMET (103)

Residue and Analytical Aspects

Phosmet has been evaluated by the JMPR several times since 1976. In New Zealand, phosmet is used at a spray concentration of 0.112% ai to control insect and scale pests on fruit, including kiwi fruit and feijoas. Residue data from supervised trials on kiwi fruit in New Zealand were supplied to the meeting.

Phosmet residues on kiwi fruit declined very slowly. The insecticide may be applied up to 7 times during the growing season. The slow rate of decline caused some accumulation of phosmet from previous applications. Most of the residues (91-94%) were in the inedible skin. No information was provided about the effect of commercial post-harvest handling on the residues. Brushing kiwi fruit in packing sheds removes some of the hairs and would be expected to reduce pesticide residues.

All the residues in the 1986 trials (maximum rate: 7 applications of 2.4 kg ai/ha) were less than 10 mg/kg on the whole fruit, while earlier trials at higher rates (7 x 3.6 kg ai/ha) produced residues up to 25 mg/kg. If the 1986 trials represent the current use pattern, current residues will be within the MRL of 15 mg/kg.

A single trial on feijoas (5 applications of 3.38 kg ai/ha) produced residues of 0.9 mg/kg at the New Zealand witholding period of 14 days. When phosmet was used at the same rate on apples similar residue levels were recorded. Feijoa is a minor crop and it is reasonable to recommend an MRL based on extrapolation and the limited data supplied to the meeting.

The meeting was unable to carry out a review (requested by the CCPR) of MRLs on maize, maize fodder, maize forage and sweet corn (corn-on-the-cob) because residue data have not yet been received.

Further Work or Information

Desirable

1.   Information on commercial practices which might affect phosmet residues on kiwi fruit, with particular reference to the brushing of kiwi fruit in packing sheds, and the proportion of the crop so treated.

2.   Information on residues on maize, maize fodder, maize forage and sweet corn (corn-on-the-cob).

4.34   PHOXIM (141)

Residue and Analytical Aspects

The 1984 JMPR reviewed data on residues in milk from cattle after dermal treatments with phoxim in The Netherlands. Residues reached their maximum concentrations of 0.05 mg/kg (average of 5 cows) during the first 24 hours after treatment and decreased in milkings 2 1/2 days later to less than the limit of determination. The 1984 JMPR had proposed that the limit of determination of 0.01 mg/kg was suitable for use as an MRL.

The meeting reviewed the data and agreed that, in view of the possibility of samples being taken for control purposes close to treatment, it would be advisable to estimate the maximum level without an implied withdrawal period of 3 days. This estimate for bulked milk was 0.05 mg/kg and the meeting recommended that this level was suitable for use as an MRL.

4.35   PROCHLORAZ (142)

Residues and Analytical Aspects

Additional information on use patterns in The Netherlands was provided together with limited residue data from supervised trials on mushrooms and stone fruits.

This additional information supports the MRLs for mushrooms and stone fruits recommended in 1985.

<u>Further work or Information</u>

<u>Required</u>

    1.   Information on nationally approved uses on citrus and papaya.

<u>Desirable</u>

    1.   Documentation that the analytical procedures separate 2,4,6-trichlorophenol from other trichlorophenols.

    2.   Validation of the analytical method for animal products determining the major trichlorphenol animal metabolites 2-(2,4,6-trichlorophenoxy)ethylurea (BTS-44770) and 2-(2,4,6-trichlorophenoxy)ethanol (BTS-3037).

4.36 PROPAMOCARB (148)

Residue and Analytical Aspects

It was noted at the 19th (1987) Session of the CCPR that data on residues in sweet peppers, which indicated the need for an MRL higher than the current 0.2 mg/kg, had been supplied to the JMPR but had not been included in the 1986 Evaluations.

It appears that the data in question are results quoted in Table 5 of the 1984 evaluation, showing residues up to 0.9 mg/kg 3-21 days after treatment with a continuous flow of nutrient film containing about 108-217 mg ai/l. The treatment gives about 0.6 l/plant. This treatment was not understood to be GAP when the data were evaluated and the estimated maximum residue level therefore did not take them into account. Such treatments reflect current GAP, however, and an increase in the MRL to 1 mg/kg is appropriate.

4.37 PROPICONAZOLE (160)

(+)-1-[2-(2,4-dichlorophenyl)-4-propyl-1,3-dioxolan-2-ylmethyl]-1H-1,2,4-triazole.

Propiconazole has not been previously evaluated by the Joint Meeting. It is a systemic fungicide with high activity against several fungal pathogens which cause a wide range of problems such as powdery mildew, rusts, and leaf spot diseases. The compound can be used for protective, curative and eradicative purposes, but the best results are achieved when it is applied while the disease is active but still in the early stages of development.

Toxicology

After oral administration to mice and rats the compound was rapidly excreted via the urine and faeces. Excretion was generally higher in the urine. No unchanged propiconazole was excreted in the urine. The compound is metabolized by oxidation of the propyl side chain, cleavage of the dioxolane ring and oxidation of the triazole and phenyl rings. The alcoholic and phenolic derivatives are excreted as sulphuric and glucuronic acid conjugates. The compound apparently induces drug metabolizing enzymes in the liver of rats and mice.

Propiconazole presented low acute toxicity in mice, rats and hamsters. It did not show any effects on reproductive performance. Maternal effects, decreased pup growth, swelling of liver hepatocytes and an increased incidence of clear-cell changes in the liver were seen at 500 ppm and above in a two-generation reproduction study in rats. In three rat and two rabbit teratogenicity studies, embryotoxic effects were found at maternally toxic dose levels, but no teratogenic effects were observed.

Subchronic administration of propiconazole to rats, rabbits and dogs revealed body weight reduction and liver changes indicative of increased metabolic activities with necrosis at relatively high doses as the main toxicological effects.

Propiconazole was not mutagenic in a battery of mutagenicity tests.

In long-term studies the same effects were found as in short-term studies. In the mouse study the highest dose (2500 ppm) was associated with a significant increase in preneoplastic and neoplastic liver changes in males only. In the two year feeding study with rats no indication of carcinogenicity was found. However, propiconazole has enzyme-inducing properties. In a separate study in rats, the effect of propiconazole on proliferative liver changes, initiated by N-nitrosodiethylamine, was very similar to that of phenobarbital.

An ADI was allocated for propiconazole, on the basis of the toxicological data in rat and dog.

The committee noted that application of propiconazole resulted in residues of triazolylalanine as well as propiconazole. The available data on propiconazole cannot be used to assess the toxicity of triazolylalanine since this does not appear to be a metabolite in animal species other than ruminants. While an ADI has been allocated for propiconazole, attention is drawn to the need for an evaluation of triazolylalanine to assess the toxicological significance of this metabolite.

Toxicological evaluation

Level causing no toxicological effect

Rat: 100 ppm in the diet, equal to 4 mg/kg bw/day
Dog: 250 ppm in the diet, equivalent to 7 mg/kg bw/day

Estimate of acceptable daily intake for man

0-0.04 mg/kg bw.

Studies which will provide information valuable in the continued evaluation of the compound

Observations in man.

Residue and Analytical Aspects

Commercial products are recommended for foliar application, by either ground or aerial equipment, usually at intervals of 14 to 28 days with a maximum of three to eight sprays per season, depending on the crop being treated. For optimum control of diseases in cereals in some conditions, prepacks or tank-mixes of propiconazole with other fungicides, such as carbendazim, chlorothalonil, tridemorph, fenpropimorph or mancozeb, are recommended.

Propiconazole is registered for use on many crops in a wide range of countries in temperate, subtropical and tropical regions. The main fields of application are in cereals (wheat, barley, oats and rye) and bananas, with lesser importance for diseases of coffee, peanuts, rice, pecan, sugarbeet, sugarcane and stone fruits; it also has uses on non-edible crops such as grassland and turf.

Extensive residue data on cereals (wheat, barley, oats, rye, rice) were provided, together with lesser amounts on grapes, bananas, stone fruits, nuts, sugar beet, sugar cane, rape seed and coffee. Apart from stone fruits and grapes treated close to harvest, residues at harvest were generally at low levels or at or about the corresponding limits of determination. The data were sufficient to enable estimates of likely maximum residue levels to be made for a number of crops and corresponding recommendations to be put forward.

The metabolism of propiconazole in plants and animals has been extensively studied using material labelled with ^{14}C in the triazole and/or the phenyl ring and the oxidative pathways of degradation involved have been elucidated satisfactorily. These studies also confirmed that residues in grapes are eliminated during processing into wine. Similar studies of degradation in soil showed half-lives of 30 to 70 days in various soil types.

Gas chromatography using a nitrogen-specific detection system was used to obtain all the residue data submitted. Such methods should be suitable for use for enforcement purposes.

Information on MRLs already in force for various crops in many countries was also available to the meeting. The available data enabled the meeting to estimate maximum residue levels for a range of commodities (see Annex I).

<u>Further Work or Information</u>

<u>Required</u>

Data on residues of propiconazole in rice from supervised trials using recommended registered rates of application and determined as the parent compound.

<u>Desirable</u>

Confirmatory data on residues in coffee beans.

4.38 PYRAZOPHOS (153)

<u>Residues and Analytical Aspects</u>

Additional information was provided to the meeting on uptake and metabolism in lactating goats, residues resulting from supervised trials and national maximum residue limits.

It has been demonstrated that orally administered ^{14}C-pyrazophos is readily absorbed in the lactating goat. Most of the dose is rapidly excreted via the kidneys in the form of metabolites; no ^{14}C-pyrazophos was detected in the urine. Levels of radioactivity in both milk and tissues were very low.

The residues data available to the meeting are in accord with those submitted to the 1985 JMPR which formed the basis for guideline levels recorded at that meeting.

Further Work or Information

Desirable

1.   Information on residues in meat from pigs and meat and eggs from poultry fed a diet containing pyrazophos.

2.   Additional information on the identities and quantities of metabolites in plants after treatment with pyrazophos.

3.   Information on the effect of processing on residues in crops.

4.39   TECNAZENE (115)

Residue and Analytical Aspects

The CCPR at its 19th (1987) Session requested the JMPR to propose MRLs for specific vegetables or restricted groups, rather than for the broad group "other vegetables" proposed at the 1978 JMPR.

The 1978 JMPR, had received and evaluated residues data from supervised trials on potatoes, glasshouse lettuce, fodder beets and Witloof chicory (sprouts). Some monitoring data had been available to that meeting, showing measurable residues in potatoes, lettuce and carrots.

Since the 1978 JMPR, no additional data have become available on GAP for tecnazene on potatoes. The post-harvest use on carrots for the control of rot during storage and transport is not considered to be current good agricultural practice.

Since the meeting knew of no other vegetables on which tecnazene may be used, and no information was available on current GAP or on residues from supervised trials on specific vegetables, the meeting concluded that the maximum residue limit for "vegetables (except chicory)" proposed at the 1978 JMPR should be withdrawn.

Further Work or Information

Required

Information on good agricultural practice in the pre- and post-harvest use of tecnazene on vegetables other than potatoes, and residues data from supervised trials on treatments according to GAP.

4.40   THIODICARB (154)

Residue and analytical aspects

The 19th (1987) Session of the CCPR requested the JMPR to re-examine the proposed MRL for sweet corn of 2 mg/kg and to review the definition of the residue.

The meeting noted that the 1985 JMPR had received a considerable amount of data on sweet corn, most of it following multi-applications of thiodicarb and sampled at 0-3 days after the last application. The analyses were carried out on the kernel plus cob and the majority of residues at day 0 (the registered interval between last application and harvest) were less than 0.1 mg/kg (14 of 20 trials). In five other trials the residues were between 0.1 and 1 mg/kg and in only one trial did residues exceed 1 mg/kg (1.3, 1.5 mg/kg). These were identified in the 1985 monographs as statistical outliers but not discarded. The meeting agreed that these data should be discarded and proposed that a new estimate of the maximum residue at 1 mg/kg (kernel plus cob) was suitable for consideration as an MRL.

The meeting also reviewed the definition of the residue and referred to Section 2.8 which deals with pesticides which are also metabolites of other pesticides. Confirming that the general approach in Section 2.8 was appropriate to the case of thiodicarb/methomyl the meeting recommended that the residue and analytical aspects of these compounds be studied in detail at a future meeting.

4.41 THIRAM (105)

Toxicology

The 1985 Meeting concluded that the assessment of all the data relating to the safety of thiram was required before the estimation of an ADI could be reconsidered.

A number of studies have been provided, but they were not considered to be adequate for evaluating the safety of thiram, because they raised more questions than they answered about many aspects of toxicology, including potential teratogenicity, reproductive toxicity, and neurotoxicity. The meeting was therefore unable, on the data available, to estimate an ADI or evaluate the safety of residues known to occur in food from current uses. A toxicological monograph was not published because of the inadequacy of the data submitted.

Residue and Analytical Aspects

The meeting noted that the regulatory method of analysis for thiram was not specific and therefore recommended that in view of the lack of an ADI the residue aspects of the compound should be reviewed as soon as possible.

4.42 TRIADIMEFON (133)

Residue and Analytical Aspects

At the 19th (1987) Session of the CCPR, it was requested that proposals for MRLs for products of animal origin should be developed as a consequence of the proposed MRLs for animal feedstuffs. This was interpreted as a request to reconsider the existing proposals for products of animal origin in the light of increased proposals for feedstuffs.

The 1979 JMPR estimated limits for triadimefon in eggs and milk (GLs at that time), the 1981 JMPR for meat, and the 1983 JMPR for poultry meat. These are all at the limit of determination, 0.1 mg/kg. Limits for grain straws and fodders were increased from 2 to 5 mg/kg by the 1984 JMPR, and for sugar beet from 0.1 to 2 mg/kg by the 1986 JMPR.

It is clear from the 1979, 1981 and 1983 evaluations that the proposals for animal products were based on feeding trials with much higher levels of triadimefon and triadimenol in the feed than the increased MRLs subsequently proposed for animal feedstuffs. There is therefore no reason to increase the proposed MRLs for meat, poultry meat, eggs or milk.

4.43 TRICHLORFON (066)

Residue and Analytical Aspects

The CCPR at its 19th (1987) Session concluded that as fresh banana pulp was not a commodity in international trade, MRLs for banana pulp should be deleted. Where possible, Codex MRLs should be established for whole bananas. The Committee understood that the residue levels occurring in the pulp would be presented as information in JMPR monographs.

The 1971 JMPR had not estimated a maximum residue level for trichlorfon in whole bananas, although data on residues in the peel were available from the trials on which the proposal for pulp had been based.

When residue data on pulp and peel are available, it is in principle possible to calculate the residue in whole bananas, since the weight of the peel is about 30% of the total weight. The meeting therefore reviewed in detail the data available to the 1971 JMPR.

The detailed data on which the 1971 evaluation was based were made available to the present meeting by the manufacturer, and were used for the calculations.

In order to accomodate some variation in the peel/pulp ratio a "worst case" calculation was made on the basis of a pulp:peel ratio of 1:1. This enabled the meeting to estimate a maximum residue level for trichlorfon in whole banana, which is recommended for use as an MRL.

4.44 VAMIDOTHION (078)

Residue and Analytical Aspects

During the 19th Session of the CCPR several countries raised objections to the proposed MRL for vamidothion in pome fruit in view of the relatively high toxicity of the compound. In order to be able to reconsider the proposal countries were invited to supply data on GAP in pome fruit to the JMPR. Only one country supplied information, from which it appears that the PHI is 30 days, whereas the PHI on which the MRL is based is 42-56 days. It was therefore decided not to change the proposal until further information on GAP is obtained.

Further Work or Information

Desirable

Information on GAP in pome fruit.

4.45 VINCLOZOLIN (159)

Residue and Analytical Aspects

In response to the requirements of the 1986 JMPR for information on GAP related to post-harvest or at-harvest uses of vinclozolin on pome and stone fruit and residues from supervised trials involving such uses, the meeting received information from two countries, Australia and the UK, in which such uses were authorized or recommended and also data on authorized applications on stone fruit close to harvest in the USA (PHI not more than 3 days). The uses mentioned are intended for the control of fungal diseases causing fruit rot during storage, transport and distribution.

The additional data received by the 1987 JMPR, together with those evaluated in 1986, enabled the meeting to estimate maximum residue levels for the post-harvest or at-harvest uses of vinclozolin on apricots, cherries and peaches, which are suitable for use as TMRLs. The data available from supervised trials on post-harvest or at-harvest applications on pome fruit and plums were insufficient to propose maximum residue limits for such uses.

Extensive information was received from New Zealand on GAP in the use of vinclozolin on various fruits and vegetables. Some data also became available on residues from supervised trials on corn salad (lambs lettuce) from The Netherlands. Since the latter data refer to the parent compound only, no new proposals could be developed for this crop. The additional data received from New Zealand confirm in general the limits established by the 1986 JMPR and no changes are needed to the limits for dewberries (including boysenberry), black currants, grapes or kiwi fruit.

The limited data on residues from supervised trials on blueberries and the similarity to other berries allow a limit to be proposed as suitable for use as a TMRL. The new data from New Zealand on residues in lettuce and tomatoes, after observing GAP, confirm the data presented at the 1986 JMPR.

The meeting expressed the opinion that definition of the residue as the parent compound only would be preferable to the present definition ("sum of vinclozolin and all metabolites containing 3,5-dichloroaniline, expressed as vinclozolin") if data became available which would allow it (see Section 3.1).

Further Work or Information

Desirable

1. Additional information on residues from supervised trials according to GAP on pome fruit (post-harvest or at-harvest).

2. Data on residues from supervised trials on crops for which GAP exists in countries involved in international trade.

5. RECOMMENDATIONS

5.1   In the interests of public health and agriculture and in view of the needs of the Codex Committee on Pesticide Residues, the meeting recommends that Joint Meetings on Pesticide Residues should continue to be held annually.

5.2   In view of the human data indicating increased incidence of bladder cancer following exposure to 4-chloro-2-toluidine, a metabolite of chlordimeform, and the occurrence of this compound as a residue in treated crops, the meeting recommends that chlordimeform should not be used where its residues, or those of its metabolite 4-chloro-2-toluidine, can arise in food.

5.3   The meeting recommends that in view of the lack of an ADI the residue aspects of thiram should be reviewed as soon as possible.

5.4   The meeting recommends that a detailed review of all aspects of the use of captan and folpet be carried out at the 1989 Joint Meeting or as soon as possible.

5.5   The meeting recommends that data on the toxicity of triazolylalanine, a metabolite of several pesticides, should be obtained and evaluated to determine its toxicological significance.

6. FUTURE WORK

The following items should be considered at the 1988 and/or 1989 meeting.

6.1   New Compounds

Compounds recommended for priority attention by the 19th or earlier Sessions of the CCPR which have not yet been evaluated.

6.1.1   For Toxicological Evaluation:

1988	1989 (tentative)
paclobutrazol	anilazine
tolylfluanid	chlorpropham
	propham
	triadimenol
	triazolylalanine*

6.1.2   For Residue Evaluation:

1988	1989 (tentative)
paclobutrazol	amilazine
tolylfluanid	chlorpropham
	propham
	triadimenol
	triazolylalanine*

6.2   **Re-evaluations**

6.2.1   **For Toxicological Evaluation:**

1988
bromide ion
butocarboxim
dimethipin (JMPR 1987)
dinocap (CCPR 1987)
ethylenethiourea (ETU)(JMPR 1986)
fenitrothion
methacrifos (JMPR 1986)
"d-phenothrin" (JMPR 1984)
vamidothion (JMPR 1985)
vinclozolin (JMPR 1986)

1989
captan
demeton-S-methyl
daminozide
endosulfan (JMPR 1985)
ethion (JMRP 1986)
folpet (JMPR 1986)
lindane
methomyl (JMPR 1986)
2-phenylphenol (JMPR 1985)
procymidone
propoxur

6.2.2   **For Residue Evalaution:**

1988
aldicarb (CCPR 1987)
bitertanol (CCPR 1987)
bromide ion (new data)
carbendazim/benomyl/thio-
phanate-methyl (JMPR 1987)
carbosulfan (JMPR 1984)
chlorothalonil (CCPR 1987)
cyhalothrin (new data)
cypermethrin (CCPR 1987)
etrimfos (JMPR 1987)
fenvalerate (CCPR 1987)
methiocarb (JMPR 1987)
permethrin (new data)
phosmet (JMPR 1987)
phoxim (CCPR 1987)
prochloraz (CCPR 1987, JMPR 1987)
thiodicarb/methomyl (JMPR 1987)

1989 (tentative)
captan (CCPR 1986)
clofentezine (JMPR 1987)
cyfluthrin (JMPR 1986)
daminozide (CCPR 1987)
endosulfan (JMRP 1985)
fenthion (CCPR) 1987)
folpet (CCPR) 1986)
metalaxyl (CCPR 1986)
procymidone (CCPR 1987)
thiram (JMPR 1987)

---

* Resulting from the 1987 JMPR

7. REFERENCES

PREVIOUS FAO AND WHO DOCUMENTS

FAO/WHO.   Principles governing consumer safety in relation to pesticide
(1962)   residues. Report of a meeting of a WHO Expert Committee on Pesticide
          Residues held jointly with the FAO Panel of Experts on the Use of
          Pesticides in Agriculture. FAO Plant Production and Protection
          Division Report, No. PL/1961/11; WHO Technical Report Series, No. 240.

FAO/WHO.   Evaluation of the toxicity of pesticide residues in food; report of a
(1964)   Joint Meeting of the FAO Committee on Pesticides in Agriculture and
          the WHO Expert Committee on Pesticide Residues. FAO Meeting Report,
          No. PL/1963/13; WHO/Food Add./23.

FAO/WHO.   Evaluation of the toxicity of pesticide residues in food. Report of
(1965a)   the Second Joint Meeting of the FAO Committee on Pesticides in
          Agriculture and the WHO Expert Committee on Pesticide Residues. FAO
          Meeting Report, No. PL/1965/10; WHO/Food Add./26.65.

FAO/WHO.   Evaluation of the toxicity of pesticide residues in food. FAO
(1965b)   Meeting Report, No. PL/1965/10/1; WHO/Food Add./27.65.

FAO/WHO.   Evaluation of the hazards to consumers resulting from the use of
(1965c)   fumigants in the protection of food. FAO Meeting Report, No.
          PL/1965/10/2; WHO/Food Add./28.65.

FAO/WHO.   Pesticide residues in food. Joint report of the FAO Working Party on
(1967a)   Pesticide Residues and the WHO Expert Committee on Pesticide
          Residues. FAO Agricultural Studies, No. 73; WHO Technical Report
          Series, No. 370.

FAO/WHO.   Evaluation of some pesticide residues in food. FAO/PL:CP/15;
(1967b)   WHO/Food Add./67.32.

FAO/WHO.   Pesticide residues. Report of the 1967 Joint Meeting of the FAO
(1968a)   Working Party and the WHO Expert Committee. FAO Meeting Report, No.
          PL:1967/M/11; WHO Technical Report Series, No. 391.

FAO/WHO.   1967 Evaluations of some pesticide residues in food. FAO/
(1968b)   PL:1967/M/11/1; WHO/Food Add./68.30.

FAO/WHO.   Pesticide residues in food. Report of the 1968 Joint Meeting of the
(1969a)   FAO Working Party of Experts on Pesticide Residues and the WHO Expert
          Committee on Pesticide Residues. FAO Agricultural Studies, No. 78;
          WHO Technical Report Series, No. 417.

FAO/WHO.   1968 Evaluation of some pesticide residues in food. FAO/
(1969b)   PL:1968/M/9/1; WHO/Food Add./69.35.

FAO/WHO.   Pesticide residues in food. Report of the 1969 Joint Meeting of the
(1970a)   FAO Working Party of Experts on Pesticide Residues and the WHO Expert
          Group on Pesticide Residues. FAO Agricultural Studies, No. 84; WHO
          Technical Report Series, No. 458.

FAO/WHO. 1969 evaluations of some pesticide residues in food. FAO/
(1970b)   PL:1969/M/17/1; WHO/Food Add./70.38.

FAO/WHO. Pesticide residues in food. Report of the 1970 Joint Meeting of the
(1971a)   FAO Working Party of Experts on Pesticide Residues and the WHO Expert
          Committee on Pesticide Residues. FAO Agricultural Studies, No. 87;
          WHO Technical Report Series, No. 474.

FAO/WHO. 1970 Evaluations of some pesticide residues in food. AGP:
(1971b)   1970/M/12/1; WHO/Food Add./71.42.

FAO/WHO. Pesticide residues in food. Report of the 1971 Joint Meeting of the
(1972a)   FAO Working Party of Experts on Pesticide Residues and the WHO Expert
          Committee on Pesticide Residues. FAO Agricultural Studies, No. 88;
          WHO Technical Report Series, No. 502.

FAO/WHO. 1971 Evaluations of some pesticide residues in food. AGP-1971/M/9/1;
(1972b)   WHO Pesticide Residues Series, No. 1.

FAO/WHO. Pesticide residues in food. Report of the 1972 Joint Meeting of the
(1973a)   FAO Working Party of Experts on Pesticide Residues and the WHO Expert
          Committee on Pesticide Residues. FAO Agricultural Studies, No. 90;
          WHO Technical Report Series, No. 525.

FAO/WHO. 1972 Evaluations of some pesticide residues in food. AGP:1972/M/9/1;
(1973b)   WHO Pesticide Residues Series, No. 2.

FAO/WHO. Pesticide residues in food. Report of the 1973 Joint Meeting of the
(1974a)   FAO Working Party of Experts on Pesticide Residues and the WHO Expert
          Committee on Pesticide Residues. FAO Agricultural Studies, No. 92;
          WHO Technical Report Series, No. 545.

FAO/WHO. 1973 Evaluations of some pesticide residues in food. FAO/AGP/1973/
(1974b)   M/9/1; WHO Pesticide Residues Series, No. 3.

FAO/WHO. Pesticide residues in food. Report of the 1974 Joint Meeting of the
(1975a)   FAO Working Party of Experts on Pesticide Residues and the WHO Expert
          Committee on Pesticide Residues. FAO Agricultural Studies, No. 97;
          WHO Technical Report Series, No. 574.

FAO/WHO: 1974 Evaluations of some pesticide residues in food. FAO/AGP/
(1975b)   1974/M/11; WHO Pesticide Residues Series, No. 4.

FAO/WHO. Pesticide residues in food. Report of the 1975 Joint Meeting of the
(1976a)   FAO Working Party of Experts on Pesticide Residues and the WHO Expert
          Committee on Pesticide Residues. FAO Plant Production and Protection
          Series, No. 1; WHO Technical Report Series, No. 592.

FAO/WHO. 1975 Evaluations of some pesticide residues in food. AGP:1975/M/13;
(1976b)   WHO Pesticide Residues Series, No. 5.

FAO/WHO. Pesticide residues in food. Report of the 1976 Joint Meeting of the
(1977a)   FAO Panel of Experts on Pesticide Residues and the Environment and
          the WHO Expert Group on Pesticide Residues. FAO Food and Nutrition
          Series, No. 9; FAO Plant Production and Protection Series, No. 8; WHO
          Technical Report Series, No. 612.

FAO/WHO. 1976 Evaluations of some pesticide residues in food. AGP:1976/M/14.
(1977b)

FAO/WHO. Pesticide residues in food - 1977. Report of the Joint Meeting of
(1978a) the FAO Panel of Experts on Pesticide Residues and Environment and
         the WHO Expert Group on Pesticide Residues. FAO Plant Production and
         Protection Paper 10 Rev.

FAO/WHO. Pesticide residues in food: 1977 evaluations. FAO Plant Pro-
(1978b) duction and Protection Paper 10 Sup.

FAO/WHO. Pesticide residues in food - 1978. Report of the Joint Meeting of
(1979a) the FAO Panel of Experts on Pesticide Residues and Environment and
         the WHO Expert Group on Pesticide Residues. FAO Plant Production and
         Protection Paper 15.

FAO/WHO. Pesticide residues in food: 1978 evaluations. FAO Plant
(1979b) Production and Protection Paper 15 Sup.

FAO/WHO. Pesticide residues in food - 1979. Report of the Joint Meeting of
(1980a) the FAO Panel of Experts on Pesticide Residues in Food and the
         Environment and the WHO Expert Group on Pesticide Residues. FAO
         Plant Production and Protection Paper 20.

FAO/WHO. Pesticide residues in food: 1979 evaluations. FAO Plant
(1980b) Production and Protection Paper 20 Sup.

FAO/WHO. Pesticide residues in food - 1980. Report of the Joint Meeting of
(1981a) the FAO Panel of Experts on Pesticide Residues in Food and the
         Environment and the WHO Expert Group on Pesticide Residues. FAO
         Plant Production and Protection Paper 26.

FAO/WHO. Pesticide residues in food: 1980 evaluations. FAO Plant
(1981b) Production and Protection Paper 26 Sup.

FAO/WHO. Pesticide residues in food - 1981. Report of the Joint Meeting of
(1982a) the FAO Panel of Experts on Pesticide Residues in Food and the
         Environment and the WHO Expert Group on Pesticide Residues. FAO
         Plant Production and Protection Paper 37.

FAO/WHO. Pesticide residues in food: 1981 evaluations. FAO Plant
(1982b) Production and Protection Paper 42.

FAO/WHO. Pesticide residues in food - 1982. Report of the Joint Meeting of
(1983a) the FAO Panel of Experts on Pesticide Residues in Food and the
         Environment and the WHO Expert Group on Pesticide Residues. FAO
         Plant Production and Protection Paper 46.

FAO/WHO. Pesticide residues in food: 1982 evaluations. FAO Plant
(1983b) Production and Protection Paper 49.

FAO/WHO. Pesticide residues in food - 1983. Report of the Joint Meeting of
(1984)   the FAO Panel of Experts on Pesticide Residues in Food and the
         Environment and the WHO Expert Group on Pesticide Residues. FAO
         Plant Production and Protection Paper 56.

FAO/WHO. Pesticide residues in food: 1983 evaluations. FAO Plant
(1985a) Production and Protection Paper 61.

FAO/WHO. Pesticide residues in food - 1984. Report of the Joint Meeting
(1985b) on Pesticide Residues. FAO Plant Production and Protection Paper 62.

FAO/WHO. Pesticide residues in food - 1984 evaluations. FAO Plant
(1985c) Production and Protection Paper 67.

FAO/WHO. Pesticide residues in food - 1985. Report of the Joint Meeting of
(1986a) the FAO Panel of Experts on Pesticide Residues in Food and the
         Environment and a WHO Expert Group on Pesticide Residues. FAO Plant
         Production and Protection Paper 68.

FAO/WHO. Pesticide residues in food - 1985 evaluations. Part I - Residues.
(1986b) FAO Plant Production and Protection Paper 72/1.

FAO/WHO. Pesticide residues in food - 1986 evaluations. Part II - Toxicology.
(1986c) FAO Plant Production and Protection Paper 72/2.

FAO/WHO. Pesticide residues in food - 1986. Report of the Joint Meeting of
(1986d) the FAO Panel of Experts on Pesticide Residues in Food and the
         Environment and a WHO Expert Group on Pesticide Residues. FAO Plant
         Production and Protection Paper 77.

FAO/WHO. Pesticide residues in food - 1986 evaluations. Part I - Residues.
(1986e) FAO Plant Production and Protection Paper 78.

CORRIGENDA TO 1986 REPORT AND RESIDUE EVALUATIONS

NOTE: (1)   Minor typographical errors are not included.

         (2)   All except the first of the corrigenda to the report are
                    given in the 1986 Evaluations, but are repeated here for
                    convenience.

REPORT:

         The following changes should be made (page numbers refer to English
edition):

1.   Page 2

         Section 2.2 Delete second para. Replace by:

                    "NOAELSs will be stated in the report for those studies
                    that are used in the overall toxicological evaluation."

2.   4.11 - CYFLUTHRIN, page 22

         Item 20 , last line: change "rapeseed to "rape seed".

3.   4.22 - FENVALERATE, page 29

         After last line of "Residue and Analytical Aspects", add:

         Further Work of Information

         Desirable

         Further information on residues of fenvalerate in Chinese
cabbage, beans (dry) and peas (dry) arising from registered uses.

4.   4.24 - GLYPHOSATE, page 32

         Para. 5, last line: change "pig and poultry meat and eggs" to "meat,
milk and eggs".

5.   4.32 - METHOPRENE, page 39

         Para. 3, first line: change "reuested clarification of the limits
for" to "requested clarification of the data on".

6.   Annex I

         2,4-D, page 59

         Previous MRL: change "except barley, maize, oats & wheat" to "except
barley, oats, rye & wheat".

7.   GLYPHOSATE, page 62

Change CCN "(157)" to "(158)".

8.   VINCLOZOLIN, page 67

Change CCN "(158)" to "(159)".

EVALUATIONS

The following changes should be made:

1.   Page 43

REFERENCES.   Add the following:

Gartrell, M.J., Craun, J.C., Podrebarac, D.S. and Gunderson, E.L. 1985. Pesticides, selected elements, and other chemicals in infant and toddler total diet samples, October 1979 - September 1980. J. Assoc. Off. Anal. Chem. 68 (6) 1163-1183

Gartrell, M.J., Craun, J.C., Podrebarac, D.S. and Gunderson, E.L. 1985. Pesticides, selected elements, and other chemicals in adult total diet samples, October 1979 - September 1980. J. Assoc. Off. Anal. Chem. 68 (6) 1184-1195

2.   Page 139

Line 6.   Change "MRLs" to "MRL"
Line 22.   Change "In plants" to "In animals"

3.   Page 183

Last line.   Change "a (T) to "an"

4.   Page 266

Headings to the Table.   Change to the following:

Commodity	CCN	Current MRLs	Recommended MRLs, mg/kg	Interval (days) 1/

5.   Page 287

Line 5 of Table.   Delete "(provisionally 10 mg/kg)"

6. <u>Pages 296, 299, 301</u>

<u>Headings to Table 2.</u> Change "Stone fruit, pre-harvest application" to "Stone fruits, pre- and post-harvest applications". (In the headings to Table 2 on other pages, "Stone fruit" should read "Stone fruits").

7. <u>Pages 341, 342, - Table 22</u>

The following changes should be made:

	AUS	BR	CDN	CH	D

<u>Page 341</u>

	AUS	BR	CDN	CH	D
004 Grapes		change 5 to 2			
004 Strawberries			change 10 to 10*		
009 Onions				Insert 0.5	Delete 0.5
013 Endive					Insert 5
013 Lambs lettuce					Insert 5

ANNEX I

ACCEPTABLE DAILY INTAKES AND RESIDUE LIMITS

PROPOSED AT THE 1987 MEETING

These figures are additional to, or amend, those recorded in Annexes of
the reports of earlier meetings. Limits recommended at meetings from 1965 to
1977 inclusive are summarized in document FAO/WHO 1978c.

This table includes maximum acceptable daily intakes (ADIs), maximum
residue limits (MRLs) and one extraneous residue limit (ERL). These terms are
defined in Annex III of the report of the 1975 meeting.

Some ADIs are temporary: this is indicated by the year in which data are
required for re-evaluation in parenthesis below the ADI.

All MRLs for compounds with temporary ADIs are necessarily temporary.
MRLs may also be temporary, irrespective of the status of the ADI, because
additional information is required for the evaluation of residues. These are
followed by the qualification (T).

If an MRL is an amendment, the previous value is also recorded, together
with the year in which the level was estimated. The absence of a figure in
the "Previous" column indicates that the recommendation is the first for the
commodity or group concerned.

The table includes the Codex Classification Numbers (CCNs) of both the
compounds and the commodities listed, to facilitate reference to the Guide to
Codex Maximum Limits for Pesticide Residues. CCNs for commodities are those
of the revised Classification. Commodities are listed in the order of the
Types" in the Classification, and within each Type in alphabetical order.
Different Types are not differentiated by sub-headings, but are separated from
one another by spaces. The Types are listed in the following order:

Type	Code
Fruits	01
Vegetables	02
Grasses	03
Nut and seeds	04
Herbs and spices	05
Mammalian products	06
Poultry products	07
Primary animal feed commodities of plant origin	11
Secondary food commodities of plant origin	12
Derived products of plant origin	13

ACCEPTABLE DAILY INTAKES (ADIs) AND MAXIMIMUM RESIDUE LIMITS (MRLs)

Pesticide (CCN) Years(s) of previous evaluation(s)	Rec. max. ADI (mg/kg bw)	Commodity		Recommended MRL or ERL (mg/kg)	
		CCN	Name	New	Previous

ACEPHATE
  (095)
1976, 1979, 1981,
1982, 1984   0.003

Residue: acephate

Remarks: temporary ADI replaced by ADI at a higher
          level. TMRLs replaced by MRLs.

BENALAXYL
  (155)
1986   0.05

Residue: benalaxyl

Remarks: as an ADI has been estimated, all GLs
          recorded in 1986 are converted to MRLs.

BITERTANOL
  (144)
1983, 1984, 1986   0.003

Residue: bitertanol

Remarks: temporary ADI replaced by ADI at a slightly
          lower level. TMRLs replaced by MRLs.

CHINOMETHIONAT
  (080)
1968 (as
oxythioquinox),   0.006
1974, 1977, 1981,
1984

Residue: chinomethionat

Remarks: as an ADI has been estimated, all GLs
          (converted from TMRLs in 1984) are
          converted to MRLs.

CHLORDIMEFORM
  (013)
1971, 1975, 1977,   Temporary ADI
1978, 1979, 1980   WITHDRAWN
1985, 1986

Remarks: temporary MRLs withdrawn, NOT converted
          to GLs.

Pesticide (CCN) Years(s) of previous evaluation(s)	Rec. max. ADI (mg/kg bw)	Commodity CCN	Name	Recommended MRL or ERL (mg/kg) New	Previous
CHLOROTHALONIL (081) 1974, 1977, 1978, 1979, 1981, 1983, 1985	0.003 (1989)		Residue: chlorothalonil Remarks: temporary ADI extended at a higher level.		

CLOFENTEZINE
  (156)
1986
	0.02	FB 0021	Currants, Black
			  Red, White
				0.1	–

		FB 0268	Gooseberry	0.1	–
		FB 0269	Grapes	0.2	–
		FB 0272	Raspberries, Red Black	0.1	–
		FS 0012	Stone fruits	0.2	0.2 (T) (1986)
		FB 0275	Strawberry	2	2 (T) (1986)
		MO 0812	Cattle, Edible offal of	0.1	0.1 (T) (1986)
		MM 0812	Cattle meat	0.05*	0.01*(T)(1986)
		ML 0812	Cattle Milk	0.01*	0.01*(T)(1986)
		PE 0112	Eggs (Poultry)	0.05*	–
		PO 0111	Poultry, Edible offal of	0.05*	–
		PM 0110	Poultry meat	0.05*	–
			Residue: clofentezine Remarks: TMRLs for citrus fruits and cucumbers, recorded in 1986, remain temporary.		

CYFLUTHRIN
  (157)
1986
	0.02		Remarks: first toxicological evaluation.
			  No GLs were recorded in 1986.		

Pesticide (CCN) Years(s) of previous evaluation(s)	Rec. max. ADI (mg/kg bw)	Commodity CCN	Commodity Name	Recommended MRL or ERL (mg/kg) New		Recommended MRL or ERL (mg/kg) Previous
DELTAMETHRIN (135) 1980, 1981, 1982, 1984, 1985, 1986	0.01	FT 0297	Fig	0.01*		-
		FT 0305	Olives	0.1		-
		VD 0561	Field peas (dry)	1	Po	-
		VD 0071	Beans (dry)	1	Po	-
		VD 0533	Lentil (dry)	1	Po	-
		GC 0080	Cereal grains	1	Po	2 (1981)
		CM 0657	Wheat bran, unprocessed	2	Po	5 (1981)
		CF 1211	Wheat flour	0.1	Po	0.5 (1980)
		CF 1212	Wheat wholemeal	0.5	Po	1 (1984)

Residue: deltamethrin

Remarks: Po indicates that the MRL covers post-harvest uses

Pesticide (CCN) Years(s) of previous evaluation(s)	Rec. max. ADI (mg/kg bw)	Commodity CCN	Commodity Name	Recommended MRL or ERL (mg/kg) New	Recommended MRL or ERL (mg/kg) Previous
DIMETHIPIN (151) 1985	0.003 (1988)	VR 0589	Potato	0.05*	0.1*(T) (1985)
		SO 0702	Sunflower seed	0.5	0.2 (T) (1985)
		MM 0095	Meat	0.02*	-
		MO 0105	Edible offal (Mammalian)	0.02*	-
		ML 0107	Milk of cattle, goats and sheep	0.02*	-
		PM 0110	Poultry meat	0.02*	-
		PO 0111	Poultry, Edible offal of	0.02*	-
		PE 0112	Eggs (Poultry)	0.02*	-
		OC 0702	Sunflower seed oil, crude	0.1	-

Residue: dimethipin

Remarks: Temporary ADI extended at the same level. All TMRLs now temporary only because the ADI remains temporary.

Pesticide (CCN) Years(s) of previous evaluation(s)	Rec. max. ADI (mg/kg bw)	Commodity CCN	Name	Recommended MRL or ERL (mg/kg) New	Previous
DIMETHOATE (027) 1965, 1966, 1967, 1970, 1973 (see formothion) 1977, 1978, 1984, 1986	0.01				
		Residue: dimethoate			
		Remarks: temporary ADI replaced by ADI at a higher level. TMRLs replaced by MRLs.			
ETHOPROPHOS (149) 1983, 1984	0.0003				
		Residue: ethoprophos			
		Remarks: as an ADI has been estimated, all GLs recorded in 1984 are converted to MRLs.			
ETHYLENETHIOUREA (108) 1974, 1977, 1986	0.002 (1987)	--	Banana (pulp)	Withdrawn	0.01* (1974)
		Residue: ethylenethiourea			
		Remarks:			
FENAMIPHOS (085) 1974, 1977, 1978	0.0005				
		Residue: sum of fenamiphos, its sulphoxide and sulphone, expressed as fenamiphos			
		Remarks: temporary ADI replaced by ADI at a slightly higher level. TMRLs replaced by MRLs.			
FLUCYTHRINATE (152) 1985	0.02	VD 0071	Beans (dry)	0.05*	0.5 in error (1985)
		VB 0041	Cabbages, Head	0.5	0.2 (1985)
		VO 1275	Sweet corn (kernels)	0.05*	-
		GC 0645	Maize	0.05*	-
		AS 0645	Maize fodder	1	-
		OC 0691	Cotton seed oil, crude	0.2	-
		OR 0691	Cotton seed oil, edible	0.2	-
		Residue : flucythrinate			
		Remarks: limits for residues in products of animal origin remain temporary.			

Pesticide (CCN) Years(s) of previous evaluation(s)	Rec. max. ADI (mg/kg bw)	Commodity CCN	Commodity Name	Recommended MRL or ERL (mg/kg) New	Recommended MRL or ERL (mg/kg) Previous
GLYPHOSATE (158) 1986	0.3	FI 0341	Kiwi fruit	0.05*	–
		VD 0541	Soya bean (dry)	5	–
		GC 0654	Wheat	10	20 (1986)
		AL 0541	Soya bean fodder	20	–
		AL 1265	Soya bean forage (green)	5	–
		CM 0657	Wheat bran, unprocessed	50	–

Residue: glyphosate

Remarks:

Pesticide (CCN) Years(s) of previous evaluation(s)	Rec. max. ADI (mg/kg bw)	Commodity CCN	Commodity Name	Recommended MRL or ERL (mg/kg) New	Recommended MRL or ERL (mg/kg) Previous
HEPTACHLOR (043) 1965, 1966, 1967, 1968, 1969, 1970, 1974, 1975	0.0005	FI 0353	Pineapple	0.01 (ERL)	0.01 (in the edible portion (ERL) (1970)

Residue: sum of heptachlor and heptachlor epoxide

Remarks:

Pesticide (CCN) Years(s) of previous evaluation(s)	Rec. max. ADI (mg/kg bw)	Commodity CCN	Commodity Name	Recommended MRL or ERL (mg/kg) New	Recommended MRL or ERL (mg/kg) Previous
MECARBAM (124) 1980, 1983, 1985, 1986	0.002	MO 0812	Cattle, Edible offal of	0.01*	0.005* (1986)
		MM 0812	Cattle meat	0.01*	0.005* (1986)

Residue: mecarbam

Remarks:

Pesticide (CCN) Years(s) of previous evaluation(s)	Rec. max. ADI (mg/kg bw)	Commodity		Recommended MRL or ERL (mg/kg)		
		CCN	Name	New	Previous	
METALAXYL	0.03	FP 0226	Apple	0.05*	0.05*	(1985)
(138)		FI 0326	Avocado	0.2	2	(1986)
1982, 1984, 1985,		FC 0001	Citrus fruits	5 Po	5 Po	(1982)
1986		FB 0269	Grapes	2	2	(1985)
		FI 0353	Pineapple (flesh)	0.05*	0.05*	(1985)
		FB 0272	Raspberries, Red, Black	0.2	-	
		FB 0275	Strawberry	0.2	0.2	(1985)
		VS 0621	Asparagus	0.05*	-	
		VB 0400	Broccoli	0.5	0.5	(1982)
		VB 0402	Brussels sprouts	0.2	0.2	(1985)
		VB 0041	Cabbages, Head	0.5	0.5	(1982)
		VR 0577	Carrot	0.1	-	
		VB 0404	Cauliflower	0.5	0.5	(1982)
		VC 0424	Cucumber	0.5	0.5	(1982)
		VC 0425	Gherkin	0.5	0.5	(1982)
		VL 0482	Lettuce, Head	2	2	(1982)
		VC 0046	Melons, except Watermelon	0.2	0.2	(1982)
		VA 0385	Onion, bulb	0.1	0.05*	(1982)
		VP 0064	Peas, shelled	0.05*	0.05*	(1982)
		VO 0051	Peppers	1	1	(1986)
		VR 0589	Potato	0.05*	0.05*	(1982)
		VD 0541	Soya bean (dry)	0.05*	0.1	(1985)
		VL 0502	Spinach	2	5	(1986)
		VC 0431	Squash, Summer	0.2	0.2	(1982)
		VR 0596	Sugar beet	0.05*	0.05*	(1982)
		VO 0448	Tomato	0.5	0.5	(1982)
		VC 0432	Watermelon	0.2	0.2	(1982)
		VC 0433	Winter Squash	0.2	0.2	(1982)
		GC 0080	Cereal grains	0.05*	0.05*	(1982)
		SB 0715	Cacao beans	0.1	-	
		SO 0691	Cotton seed	0.05*	0.05*	(1985)
		SO 0697	Peanut	0.1	-	
		SO 0702	Sunflower seed	0.05*	0.05*	(1982)
		DH 1100	Hops, dry	10	10	(1982)

<u>Residue</u>: metalaxyl. Note change.

<u>Remarks</u>: All previous MRLs were based on the previous residue definition.
Po indicates that the MRL covers post-harvest uses. The list includes all crops currently covered by MRLs.

Pesticide (CCN) Years(s) of previous evaluation(s)	Rec. max. ADI (mg/kg bw)	Commodity CCN	Name	Recommended MRL or ERL (mg/kg) New	Previous
METHIOCARB (132) 1981, 1983, 1985, 1986	0.001	FC 0001	Citrus fruits	Withdrawn	0.05* (1981)
		FS 0014	Plums	Withdrawn	1 (1981)
		FB 0275	Strawberry	Withdrawn	0.05* (1981)
		VB 0400	Broccoli	0.2	0.2 (T) (1981)
		VB 0402	Brussels sprouts	0.2	0.2 (T) (1981)
		VB 0041	Cabbages, Head	0.2	0.2 (T) (1981)
		VB 0404	Cauliflower	0.2	0.2 (T) (1981)
		VP 0526	Common bean	Withdrawn	0.2 (T) for beans, snap and lima
		VL 0482	Lettuce, Head	0.2)	0.2 (T) for
		VL 0483	Lettuce, Leaf	0.2)	lettuce (1981)
		VR 0591	Radish, Japanese	Withdrawn	5 for Chinese radish (1983)
		VR 0596	Sugar beet	Withdrawn	0.05* (1983)
		VO 0447	Sweet corn (corn-on-the-cob)	Withdrawn	0.05*
		VO 0448	Tomato	Withdrawn	0.2
		GC 0080	Cereal grains	0.05*	-
		--	Rice (in the husk)	Withdrawn	0.5 (1981)
		GC 0651	Sorghum	Withdrawn	5 (1983)
		SO 0495	Rape seed	0.05*	-

Residue: sum of methiocarb, its sulphoxide and its sulphone

Remarks: Recommendation for cereal grains replaces separate recommendations for maize (at the same level), rice and sorghum.

Pesticide (CCN) Years(s) of previous evaluation(s)	Rec. max. ADI (mg/kg bw)	Commodity CCN	Name	Recommended MRL or ERL (mg/kg) New	Previous
METHOMYL (094) 1975, 1976, 1977, 1978, 1986	0.01 (1988)	GC 0647	Oats	0.2	0.1 (1975)
		GC 0654	Wheat	0.2	0.1 (1975)
		DH 1100	Hops, dry	2	1 (1976)

Residue: methomyl

Remarks:

Pesticide (CCN) Years(s) of previous evaluation(s)	Rec. max. ADI (mg/kg bw)	Commodity		Recommended MRL or ERL (mg/kg)	
		CCN	Name	New	Previous

Pesticide (CCN) Years(s) of previous evaluation(s)	Rec. max. ADI (mg/kg bw)	CCN	Name	New	Previous
METHOPRENE (147) 1984, 1986	0.1				

Residue: methoprene

Remarks: temporary ADI replaced by ADI at slightly higher level. TMRLs replaced by MRLs.

PERMETHRIN (120) 1979, 1980, 1981, 1982, 1983, 1984, 1985, 1986	0.05*				

Residue: permethrin (sum of isomers)

Remarks: *Applies to the nominal 40% cis-, 60% trans- and 25% cis-, 75% trans- materials only. Previously applied to 40% cis-, 60% trans- only.

PHOSMET (103) 1976, 1977, (corr. to 1976), 1978, 1979, 1981, 1984, 1985, 1986	0.02	FI 0335	Feijoa	2	-

Residue: sum of phosmet and its oxygen analogue

Remarks:

PHOXIM (141) 1982, 1983, 1984, 1986	0.001	ML 0106	Milks	0.05	0.01* (1984)

Residue: phoxim

Remarks:

PROPAMOCARB (148) 1984, 1986	0.1	VO 0445	Peppers, sweet	1	0.2 for peppers (1984)

Residue: propamocarb (base)

Remarks:

Pesticide (CCN) Years(s) of previous evaluation(s)	Rec. max. ADI (mg/kg bw)	Commodity		Recommended MRL or ERL (mg/kg)	
		CCN	Name	New	Previous
PROPICONAZOLE (160)	0.04	FI 0327	Banana	0.1	-
		FB 0269	Grapes	0.5	-
		FI 0345	Mango	0.05	-
		FS 0012	Stone fruits	1	-
		VR 0596	Sugar beet	0.05	-
		GC 0080	Cereal grains (except rice)	0.1	-
		GS 0659	Sugar cane	0.05	-
		TN 0660	Almonds	0.05	-
		SB 0716	Coffee beans	0.1	-
		SO 0697	Peanut	0.05	-
		SO 0703	Peanut, whole	0.1	-
		TN 0672	Pecan	0.05	-
		SO 0495	Rape seed	0.1	-
		AV 0596	Sugar beet leaves or tops	0.1	-
		MM 0095	Meat	0.05*	-
		MO 0105	Edible offal (Mammalian)	0.05	-
		ML 0106	Milks	0.01*	-
		PM 0110	Poultry meat	0.05*	-
		PE 0112	Eggs (Poultry)	0.05*	-

Residue: propiconazole

Remarks:

Pesticide (CCN) Years(s) of previous evaluation(s)	Rec. max. ADI (mg/kg bw)	Commodity		Recommended MRL or ERL (mg/kg)	
		CCN	Name	New	Previous
TECNAZENE (115) 1974, 1978, 1981, 1983		--	Vegetables (except chicory)	Withdrawn	0.1 (1978)

Residue: tecnazene

Remarks:

Pesticide (CCN) Years(s) of previous evaluation(s)	Rec. max. ADI (mg/kg bw)	Commodity		Recommended MRL or ERL (mg/kg)	
		CCN	Name	New	Previous
THIODICARB (154) 1985, 1986	0.03	VO 0447	Sweet corn (corn-on-the-cob)	1	2 (1985)

Residue: sum of thiodicarb, methomyl and methyl hydroxythioacetimidate ("methomyl oxime"), expressed as thiodicarb

Remarks:

Pesticide (CCN) Years(s) of previous evaluation(s)	Rec. max. ADI (mg/kg bw)	Commodity		Recommended MRL or ERL (mg/kg)	
		CCN	Name	New	Previous
TRICHLORFON (066) 1971, 1975, 1978	0.01	FI 0327	Banana	1	–

<u>Residue:</u> trichlorfon

<u>Remarks:</u>

Pesticide (CCN) Years(s) of previous evaluation(s)	Rec. max. ADI (mg/kg bw)	Commodity		Recommended MRL or ERL (mg/kg)	
		CCN	Name	New	Previous
VINCLOZOLIN (159) 1986	0.04 (1987)	FS 0240	Apricot	5 Po	–
		FS 0013	Cherries	5 Po	3 (1986)
		FS 0247	Peach	5 Po	2 (1986)
		FB 0020	Blueberries	5	–

<u>Residue:</u> sum of vinclozolin and all metabolites containing 3,5-dichloroaniline, expressed as vinclozoline

<u>Remarks:</u> Po indicates that the TMRL covers post-harvest uses

ANNEX II

FAO/WHO JOINT MEETING ON PESTICIDE RESIDUES

INDEX OF REPORTS AND EVALUATIONS

ACEPHATE   1976 (T,R)*, 1979 (R), 1981 (R), 1982 (T), 1984 (T,R), 1987 (T)

ACRYLONITRILE   1965 (T,R)

ALDICARB   1979 (T,R), 1982 (T,R), 1985 (R)

ALDRIN   1965 (T), 1966 (T,R), 1967 (R), 1974 (R), 1975 (R), 1977 (T)

ALLETHRIN   1965 (T,R)

AMINOCARB   1978 (T,R), 1979 (T,R)

AMITRAZ   1980 (T,R), 1983 (R), 1984 (T,R), 1985 (R), 1986 (R)

AMITROLE   1974 (T,R), 1977 (T)

AZINPHOS-ETHYL   1973 (T,R), 1983 (R)

AZINPHOS-METHYL   1965 (T), 1968 (T,R), 1972 (R), 1973 (T), 1974 (R)

AZOCYCLOTIN   1979 (R), 1981 (T), 1982 (R), 1983 (R), 1985 (R)

BENALAXYL   1986 (R), 1987 (T)

BENDIOCARB   1982 (T,R), 1984 (T,R)

BENOMYL   1973 (R), 1975 (T,R), 1978 (T,R), 1983 (T,R)

BHC (technical)   1965 (T), 1968 (T,R), 1973 (T,R) (see also lindane)

BINAPACRYL   1969 (T,R), 1974 (R), 1982 (T), 1984 (R), 1985 (T,R)

BIORESMETHRIN   1975 (R), 1976 (T,R)

BIPHENYL   see diphenyl

BITERTANOL   1983 (T), 1984 (R), 1986 (R), 1987 (T)

BROMIDE ION   1968 (R), 1969 (T,R), 1971 (R), 1979 (R), 1981 (R), 1983 (R)

BROMOMETHANE   1965 (T,R), 1966 (T,R), 1967 (R), 1968 (T,R), 1971 (R),
                   1979 (R), 1985 (R)

---

* T = Toxicology
  R = Residues and Analytical Aspects

BROMOPHOS	1972 (T,R), 1975 (R), 1977 (T,R), 1982 (R), 1984 (R), 1985 (R)
BROMOPHOS-ETHYL	1972 (T,R), 1975 (T,R), 1977 (R)
BROMOPROPYLATE	1973 (T,R)
BUTOCARBOXIM	1983 (R), 1984 (T), 1985 (T), 1986 (R)
sec-BUTYLAMINE	1975 (T,R), 1977 (R), 1978 (T,R), 1979 (R), 1980 (R), 1981 (T), 1984 (T,R) (withdrawal of TADI, but no T evaluation)
CAMPHECHLOR	1968 (T,R), 1973 (T,R)
CAPTAFOL	1969 (T,R), 1973 (T,R), 1974 (R), 1976 (R), 1977 (T,R), 1982 (T), 1985 (T,R), 1986 (T) (correction to 1985 report)
CAPTAN	1965 (T), 1969 (T,R), 1973 (T), 1974 (R), 1977 (T,R), 1978 (T,R), 1980 (R), 1982 (T), 1984 (T,R), 1986 (R), 1987 (R)
CARBARYL	1965 (T), 1966 (T,R), 1967 (T,R), 1968 (R), 1969 (T,R), 1970 (R), 1973 (T,R), 1975 (R), 1976 (R), 1977 (R), 1979 (R), 1984 (R)
CARBENDAZIM	1973 (T,R), 1976 (R), 1977 (T), 1978 (R), 1983 (T,R), 1985 (T,R), 1987 (R)
CARBOFURAN	1976 (T,R), 1979 (T,R), 1980 (T), 1982 (T)
CARBON DISULPHIDE	1965 (T,R), 1967 (R), 1968 (R), 1971 (R), 1985 (R)
CARBON TETRACHLORIDE	1965 (T,R), 1967 (R), 1968 (T,R), 1971 (R), 1979 (R), 1985 (R)
CARBOPHENOTHION	1972 (T,R), 1976 (T,R), 1977 (T,R), 1979 (T,R), 1980 (T,R), 1983 (R)
CARBOSULFAN	1984 (T,R), 1986 (T)
CARTAP	1976 (T,R), 1978 (T,R)
CHINOMETHIONAT	1968 (T,R) (as oxythioquinox), 1974 (T,R), 1977 (T,R), 1981 (T,R), 1983 (R), 1984 (T,R), 1987 (T)
CHLORBENSIDE	1965 (T)
CHLORDANE	1965 (T), 1967 (T,R), 1969 (R), 1970 (T,R), 1972 (R), 1974 (R), 1977 (T,R), 1982 (T), 1984 (T,R), 1986 (T)
CHLORDIMEFORM	1971 (T,R), 1975 (T,R), 1977 (T), 1978 (T,R), 1979 (T), 1980 (T), 1985 (T), 1986 (R), 1987 (T)
CHLORFENSON	1965 (T)
CHLORFENVINPHOS	1971 (T,R), 1984 (R)
CHLORMEQUAT	1970 (T,R), 1972 (T,R), 1976 (R), 1985 (R)
CHLOROBENZILATE	1965 (T), 1968 (T,R), 1972 (R), 1975 (R), 1977 (R), 1980 (T)

CHLOROPICRIN   1965 (T,R)

CHLOROPROPYLATE   1968 (T,R), 1972 (R)

CHLOROTHALONIL   1974 (T,R), 1977 (T,R), 1978 (R), 1979 (T,R), 1981 (T,R),
                           1983 (T,R), 1984 (T) (corrections to 1983 report), 1985 (T,R),
                           1987 (T)

CHLORPROPHAM   1965 (T)

CHLORPYRIFOS   1972 (T,R), 1974 (R), 1975 (R), 1977 (T,R), 1981 (R), 1982
                           (T,R), 1983 (R)

CHLORPYRIFOS-METHYL   1975 (T,R), 1976 (R) (Annex I only), 1979 (R)

CHLORTHION   1965 (T)

CLOFENTEZINE   1986 (T,R), 1987 (R)

COUMAPHOS   1968 (T,R), 1972 (R), 1975 (R), 1978 (R), 1980 (T,R), 1983 (R),
                           1987 (T)

CRUFOMATE   1968 (T,R), 1972 (R)

CYANOFENPHOS   1975 (T,R), 1978 (T) (ADI extended, but no evaluation),
                           1980 (T), 1982 (R), 1983 (T)

CYFLUTHRIN   1986 (R), 1987 (T)

CYHALOTHRIN   1984 (T,R), 1986 (R)

CYHEXATIN (TRICYCLO=   1970 (T,R), 1973 (T,R), 1974 (R), 1975 (R), 1977 (T),
HEXYLTIN HYDROXIDE   1978 (T,R), 1980 (T), 1981 (T), 1982 (R), 1983 (R), 1985 (R)
'70 & '73)

CYPERMETHRIN   1979 (T,R), 1981 (T,R), 1982 (R), 1983 (R), 1984 (R), 1985 (R),
                           1986 (R), 1987(R; correction to 1986 report)

2,4-D   1970 (T,R), 1971 (T,R), 1974 (T,R), 1975 (T,R), 1980 (R),
                           1985 (R), 1986 (R), 1987(R; correction to 1986 report)

DAMINOZIDE   1977 (T,R), 1983 (T)

DDT   1965 (T), 1966 (T,R), 1967 (T,R), 1968 (T,R), 1969 (T,R),
                           1978, (R) 1979 (T), 1980 (T), 1983 (T), 1984 (T)

DELTAMETHRIN   1980 (T,R), 1981 (T,R), 1982 (T,R), 1984 (R), 1985 (R),
                           1986 (R), 1987 (R)

DEMETON   1965 (T), 1967 (R), 1975 (R), 1982 (T)

DEMETON-S-METHYL   1965 (T), 1967 (T), 1968 (R), 1973 (T,R), 1979 (R), 1982 (T),
and related compounds   1984 (T,R)

DIALIFOS   1976 (T,R), 1982 (T), 1985 (R)

DIAZINON   1965 (T), 1966 (T), 1967 (R), 1968 (T,R), 1970 (T,R), 1975 (R),
                         1979 (R)

1,2-DIBROMOETHANE   1965 (T,R), 1966 (T,R), 1967 (R), 1968 (R), 1971 (R), 1979 (R),
                         1985 (R)

DICHLOFLUANID   1969 (T,R), 1974 (T,R), 1977 (T,R), 1979 (T,R), 1981 (R), 1982
                         (R), 1983 (T,R), 1985 (R)

1,2-DICHLOROETHANE   1965 (T,R), 1967 (R), 1971 (R), 1979 (R), 1985 (R)

DICHLORVOS   1965 (T,R), 1966 (T,R), 1967 (T,R), 1969 (R), 1970 (T,R),
                         1974 (R), 1977 (T)

DICLORAN   1974 (T,R), 1977 (T,R)

DICOFOL   1968 (T,R), 1970 (R), 1974 (R)

DIELDRIN   1965 (T), 1966 (T,R), 1967 (T,R), 1968 (R), 1969 (R),
                         1970 (T,R), 1974 (R), 1975 (R), 1977 (T)

DIFLUBENZURON   1981 (T,R), 1983 (R), 1984 (T,R), 1985 (T,R)

DIMETHIPIN   1985 (T,R), 1987 (T,R)

DIMETHOATE   1965 (T), 1966 (T), 1967 (T,R), 1970 (R), 1973 (R) (in
                         evaluation of formathion), 1977 (R), 1978 (R), 1983 (R)
                         1984, (T,R) 1986 (R), 1987 (T,R)

DIMETHRIN   1965 (T)

DINOCAP   1969 (T,R), 1974 (T,R)

DIOXATHION   1968 (T,R), 1972 (R)

DIPHENYL   1966 (T,R), 1967 (T)

DIPHENYLAMINE   1969 (T,R), 1976 (T,R), 1979 (R), 1982 (T), 1984 (T,R)

DIQUAT   1970 (T,R), 1972 (T,R), 1976 (R), 1977 (T,R), 1978 (R)

DISULFOTON   1973 (T,R), 1975 (T,R), 1979 (R), 1981 (R), 1984 (R)

DITHIOCARBAMATE   1965 (T), 1967 (T,R), 1970 (T,R), 1974 (T,R), 1977 (T,R),
  FUNGICIDES   1980 (T,R), 1983 (R) (propineb and thiram), 1984 (R)
                         (propineb), 1985 (R), 1987 (T)

DNOC   1965 (T)

DODINE   1974 (T,R), 1976 (T,R), 1977 (R)

EDIFENPHOS   1976 (T,R), 1979 (T,R), 1981 (T,R)

ENDOSULFAN   1965 (T), 1967 (T,R), 1968 (T,R), 1971 (R), 1974 (R), 1975 (R),
                         1982 (T), 1985 (T,R)

ENDRIN   1965 (T), 1970 (T,R), 1974 (R), 1975 (R)

ETHEPHON   1977 (T,R), 1978 (T,R), 1983 (R), 1985 (R)

ETHIOFENCARB   1977 (T,R), 1978 (R), 1981 (R), 1982 (T,R), 1983 (R)

ETHION   1968 (T,R), 1969 (R), 1970 (R), 1972 (T,R), 1975 (R), 1982 (T),
                            1983 (R), 1985 (T), 1986 (T)

ETHOPROPHOS   1983 (T), 1984 (R), 1987 (T)

ETHOXYQUIN   1969 (T,R)

ETHYLENE DIBROMIDE   see 1,2-dibromoethane

ETHYLENE DICHLORIDE   see 1,2-dichloroethane

ETHYLENE OXIDE   1965 (T,R), 1968 (T,R), 1971 (R)

ETHYLENETHIOUREA (ETU)   1974 (R), 1977 (T,R), 1986 (T,R), 1987 (R)

ETRIMFOS   1980 (T,R), 1982 (T), 1986 (T,R), 1987 (R)

FENAMIPHOS   1974 (T,R), 1977 (R), 1978 (R), 1980 (R), 1985 (T), 1987 (T)

FENBUTATIN OXIDE   1977 (T,R), 1979 (R)

FENCHLORPHOS   1968 (T,R), 1972 (R), 1983 (R)

FENITROTHION   1969 (T,R), 1974 (T,R), 1976 (R), 1977 (T,R), 1979 (R),
                            1982, (T) 1983 (R), 1984 (T,R), 1986 (T,R), 1987 (R)

FENSULFOTHION   1972 (T,R), 1982 (T), 1983 (R)

FENTHION   1971 (T,R), 1975 (T,R), 1977 (R), 1978 (T,R), 1979 (T),
                            1980, (T) 1983 (R)

FENTIN compounds   1965 (T), 1970 (T,R), 1972 (R), 1986 (R)

FENVALERATE   1979 (T,R), 1981 (T,R), 1982 (T), 1984 (T,R), 1985 (R),
                            1986 (T,R), 1987 (R)

FERBAM   see dithiocarbamate fungicides, 1965 (T), 1967 (T,R)

FLUCYTHRINATE   1985 (T,R), 1987 (R)

FOLPET   1969 (T,R), 1973 (T), 1974 (R), 1982 (T), 1984 (T,R), 1986 (T),
                            1987 (R)

FORMOTHION   1969 (T,R), 1972 (R), 1973 (T,R), 1978 (R)

GLYPHOSATE   1986 (T,R), 1987 (R)

GUAZATINE   1978 (T.R), 1980 (R)

HEPTACHLOR   1965 (T), 1966 (T,R), 1967 (R), 1968 (R), 1969 (R), 1970 (T,R),
                            1974 (R), 1975 (R), 1977 (R), 1987 (R)

HEXACHLOROBENZENE	1969 (T,R), 1973 (T,R), 1974 (T,R), 1978 (T), 1985 (R)
HYDROGEN CYANIDE	1965 (T,R)
HYDROGEN PHOSPHIDE	1965 (T,R), 1966 (T,R), 1967 (R), 1969 (R), 1971 (R)
IMAZALIL	1977 (T,R), 1980 (T,R), 1984 (T,R), 1985 (T,R), 1986 (T)
IPRODIONE	1977 (T,R), 1980 (R)
ISOFENPHOS	1981 (T,R), 1982 (T,R), 1984 (R), 1985 (R), 1986 (T,R)
LEAD ARSENATE	1965 (T), 1968 (T,R)
LEPTOPHOS	1974 (T,R), 1975 (T,R), 1978 (T,R)
LINDANE	1965 (T), 1966 (T,R), 1967 (R), 1968 (R), 1969 (R), 1970 (T,R) (published as Annex VI to 1971 evaluations), 1973 (T,R), 1974 (R), 1975 (R), 1977 (T,R), 1978 (R), 1979 (R)
MALATHION	1965 (T), 1966 (T,R), 1967 (R) (correction to 1966), 1968 (R), 1969 (R), 1970 (R), 1973 (R), 1975 (R), 1977 (R), 1984 (R)
MALEIC HYDRAZIDE	1976 (T,R), 1977 (T,R), 1980 (T), 1984 (T,R)
MANEB	see dithiocarbamate fungicides, 1965 (T), 1967 (T,R), 1987 (T)
MANCOZEB	1967 (T,R), 1970 (T,R), 1974 (R), 1977 (R), 1980 (T,R)
MECARBAM	1980 (T,R), 1983 (T,R), 1985 (T,R), 1986 (T,R), 1987 (R)
METALAXYL	1982 (T,R), 1984 (R), 1985 (R), 1986 (R), 1987 (R)
METHACRIFOS	1980 (T,R), 1982 (T), 1986 (T)
METHAMIDOPHOS	1976 (T,R), 1979 (R), 1981 (R), 1982 (T), 1984 (R), 1985 (T)
METHIDATHION	1972 (T,R), 1975 (T,R), 1979 (R)
METHIOCARB	1981 (T,R), 1983 (T,R), 1984 (T), 1985 (T), 1986 (R), 1987 (T,R)
METHOMYL	1975 (R), 1976 (R), 1977 (R), 1978 (R), 1986 (T,R), 1987 (R)
METHOPRENE	1984 (T,R), 1986 (R), 1987 (T)
METHOXYCHLOR	1965 (T), 1977 (T)
METHYL BROMIDE	see bromomethane
MEVINPHOS	1965 (T), 1972 (T,R)
MGK 264	1967 (T,R)
MONOCROTOPHOS	1972 (T,R), 1975 (T,R)
NABAM	see dithiocarbamate fungicides, 1965 (T), 1976 (T,R)

NITROFEN	1983 (T,R)
OMETHOATE	1971 (T,R), 1975 (T,R), 1978 (T,R), 1979 (T), 1981 (T,R), 1984 (R), 1985 (T), 1986 (R), 1987 (R)
ORGANOMERCURY	1965 (T), 1966 (T,R), 1967 (T,R)
OXAMYL	1980 (T,R), 1983 (R), 1984 (T), 1985 (T,R), 1986 (R)
OXYDEMETON-METHYL	1965 (T), 1967 (T), 1968 (R), 1973 (T,R), 1984 (T)
OXYTHIOQUINOX	see chinomethionat
PARAQUAT	1970 (T,R), 1972 (T,R), 1976 (T,R), 1978 (R), 1981 (R), 1982 (T), 1985 (T), 1986 (T)
PARATHION	1965 (T), 1967 (T,R), 1969 (R), 1970 (R), 1984 (R)
PARATHION-METHYL	1965 (T), 1968 (T,R), 1972 (R), 1975 (T,R), 1978 (T,R), 1979 (T), 1980 (T), 1982 (T), 1984 (T,R)
PERMETHRIN	1979 (T,R), 1980 (R), 1981 (T,R), 1982 (R), 1983 (R), 1984 (R), 1985 (R), 1986 (T,R), 1987 (T)
2-PHENYLPHENOL	1969 (T,R), 1975 (R), 1983 (T), 1985 (T,R)
PHENOTHRIN	1979 (R), 1980 (T,R), 1982 (T), 1984 (T), 1987 (R)
PHENTHOATE	1980 (T,R), 1981 (R), 1984 (T)
PHORATE	1977 (T,R), 1982 (T), 1983 (T), 1984 (R), 1985 (T)
PHOSALONE	1972 (T,R), 1975 (R), 1976 (R)
PHOSMET	1976 (R), 1977 (R) (correction to 1976 evaluation), 1978 (T,R), 1979 (T,R), 1981 (R), 1984 (R), 1985 (R), 1986 (R), 1987 (R)
PHOSPHINE	see hydrogen phosphide
PHOSPHAMIDON	1965 (T), 1966 (T), 1968 (T,R), 1969 (R), 1972 (R), 1974 (R), 1982 (T), 1985 (T), 1986 (T)
PHOXIM	1982 (T), 1983 (R), 1984 (T,R), 1986 (R), 1987 (R)
PIPERONYL BUTOXIDE	1965 (T,R), 1966 (T,R), 1967 (R), 1969 (R), 1972 (T,R)
PIRIMICARB	1976 (T,R), 1978 (T,R), 1979 (R), 1981 (T,R), 1982 (T), 1985 (R)
PIRIMIPHOS-METHYL	1974 (T,R), 1976 (T,R), 1977 (R), 1979 (R), 1983 (R), 1985 (R)
PROCHLORAZ	1983 (T,R), 1985 (T), 1987 (R)
PROCYMIDONE	1981 (R), 1982 (T)
PROPAMOCARB	1984 (T,R), 1986 (T,R), 1987 (R)
PROPARGITE	1977 (T,R), 1978 (R), 1979 (R), 1980 (T,R), 1982 (T,R)

PROPHAM	1965 (T)
PROPICONAZOLE	1987 (T,R)
PROPINEB	1977 (T,R), 1980 (T), 1983 (T), 1984 (R), 1985 (T,R)
PROPOXUR	1973 (T,R), 1977 (R), 1981 (R), 1983 (R)
PYRAZOPHOS	1985 (T,R), 1987 (R)
PYRETHRINS	1965 (T), 1966 (T,R), 1967 (R), 1968 (R), 1969 (R), 1970 (T), 1972 (T,R), 1974 (R)
QUINTOZENE	1969 (T,R), 1973 (T,R), 1974 (R), 1975 (T,R), 1976 (R) (Annex I, correction to 1975), 1977 (T,R)
2,4,5-T	1970 (T,R), 1979 (T,R), 1981 (T)
TECNAZENE	1974 (T,R), 1978 (T,R), 1981 (R), 1983 (T), 1987 (R)
THIABENDAZOLE	1970 (T,R), 1971 (R), 1972 (R), 1975 (R), 1977 (T,R), 1979 (R), 1981 (R)
THIODICARB	1985 (T,R), 1986 (T), 1987 (R)
THIOMETON	1969 (T,R), 1973 (T,R), 1976 (R), 1979 (T,R)
THIOPHANATE-METHYL	1973 (T,R), 1975 (T,R), 1977 (T), 1978 (R)
THIRAM	see dithiocarbamate fungicides, 1965 (T), 1976 (T,R), 1977 (T), 1983 (T), 1984 (R), 1985 (T,R), 1987 (T)
TOXAPHENE	see camphechlor
TRIADIMEFON	1979 (R), 1981 (T,R), 1983 (T,R), 1984 (R), 1985 (T,R), 1986 (R), 1987 (R)
TRIAZOPHOS	1982 (T), 1983 (R), 1984 (R, corrections to 1983 report), 1986 (T,R)
TRICHLORFON	1971 (T,R), 1975 (T,R), 1978 (T,R), 1987 (R)
TRICHLORONAT	1971 (T,R)
TRICHLOROETHYLENE	1968 (R)
TRIFORINE	1977 (T,R), 1978 (T,R)
TRIPHENYLTIN COMPOUNDS	see fentin compounds
VAMIDOTHION	1973 (T,R), 1982 (T), 1985 (T,R), 1987 (R)
VINCLOZOLIN	1986 (T,R), 1987 (R)
ZINEB	see dithiocarbamate fungicides, 1965 (T), 1967 (T,R)
ZIRAM	see dithiocarbamate fungicides, 1965 (T), 1967 (T,R)